ALWAYS DAY ONE

ALWAYS DAY ONE

How the Tech Titans Plan to Stay on Top Forever

ALEX KANTROWITZ

PORTFOLIO / PENGUIN

PORTFOLIO / PENGUIN
An imprint of Penguin Random House LLC
Penguinrandomhouse.com

Most Portfolio books are available at a discount when purchased in quantity for sales promotions or corporate use. Special editions, which include personalized covers, excerpts, and corporate imprints, can be created when purchased in large quantities. For more information, please call (212) 572-2232 or e-mail specialmarkets@penguinrandomhouse.com. Your local bookstore can also assist with discounted bulk purchases using the Penguin Random House corporate Business-to-Business program. For assistance in locating a participating retailer, e-mail B2B@penguinrandomhouse.com.

LIBRARY OF CONGRESS CATALOGING-IN-PUBLICATION DATA
Names: Kantrowitz, Alex, author.
Title: Always day one / Alex Kantrowitz.
Description: [New York] : Portfolio/Penguin, [2020] |
Includes bibliographical references and index.
Identifiers: LCCN 2019054201 (print) | LCCN 2019054202 (ebook) |
ISBN 9780593083482 (hardcover) | ISBN 9780593083499 (ebook)
Subjects: LCSH: High technology industries—Management—Case studies. |
Internet industry—Management—Case studies.
Classification: LCC HD62.37 .K36 2020 (print) | LCC HD62.37 (ebook) |
DDC 338.7/6004—dc23
LC record available at https://lccn.loc.gov/2019054201
LC ebook record available at https://lccn.loc.gov/2019054202

Printed in the United States of America
1 3 5 7 9 10 8 6 4 2

Book design by Daniel Lagin

CONTENTS

To everyone out there trying to make it

PREFACE

THE ZUCKERBERG ENCOUNTER

In February 2017, Mark Zuckerberg summoned me to his Menlo Park, California, headquarters for a meeting. It was my first time sitting down with the Facebook CEO, and it didn't go as anticipated.

His company, per usual, was enveloped in controversy. Pushing hard to grow its products but reluctant to moderate them, it had allowed them to fill with misinformation, sensationalism, and violent imagery. Zuckerberg seemed ready to talk about it, and I was eager to listen.

Facebook's main office—a vast, open, concrete slab of a structure—is an unnerving place to walk into. It has nine lobbies, two layers of security to enter, and guards demanding you sign a nondisclosure agreement before taking another step. Once inside, I made my way to a glass-walled conference room, smack in the middle of everything, where Zuckerberg holds his meetings. And after he finished a conversation with his chief operating officer, Sheryl Sandberg, he ushered me in, along with my editor Mat Honan, for a chat in full view of anyone walking by.

Zuckerberg had been hard at work on his "Manifesto," a fifty-seven-hundred-word exposition outlining Facebook's response to the troubling content, and its role in its users' lives more broadly. Coming in, I expected the typical CEO briefing: a lecture followed by limited time for questions. But after a brief overview, Zuckerberg started asking for feedback. "What are the things we talked about that you feel didn't come through in the text?" he said. "What's missing?"

As I answered, Zuckerberg listened intently. His posture didn't shift. His focus didn't break. And his reactions—first a gentle dispute of my suggestion that Facebook talk more about its power, and then an acknowledgment—made clear he wasn't asking just for show. I had never seen a CEO do this before, let alone one known for his obstinacy. It felt different, and worthy of investigation.

After our meeting, I asked everyone I could about Zuckerberg's unusual appeal for feedback. Is this normal? Has he ever asked you? Following several inquiries, I had my answer: the request was simply a glimpse into the way he runs Facebook. Zuckerberg has built feedback into Facebook's very fiber. Major meetings end with requests for it. Posters around Facebook's offices say FEEDBACK IS A GIFT. And nobody in the company is above it, not even Zuckerberg himself.

As a tech reporter in Silicon Valley, I've watched the tech giants' unconventional run of dominance from the front row. Instead of following the typical corporate life cycle—grow, slow, stumble, and ossify—Apple, Amazon, Facebook, Google, and Microsoft have only become more powerful as they've aged. And perhaps except for Apple (more on that later), they're showing little sign of letting up.

Amid this run, I've been struck by these companies' uncommon internal practices. After countless interviews with chief executives, for instance, I had been convinced the world's top CEOs were natural sellers, people who used the force of their personalities to rally others around their visions. But look at Zuckerberg and his counterparts—Jeff Bezos at Amazon, Sundar Pichai at Google, Satya Nadella at Microsoft—and you'll see trained engineers more eager to facilitate than to dictate. Instead of answers, they have questions. Instead of pitching, they listen and learn.

Following that meeting in Menlo Park, I began digging into the tech giants' inner workings more broadly—looking at their leadership practices, their cultures, their technology, and their processes—wondering if there was a link between their success and the unique way they operate. As common patterns emerged, that link became impossible to deny. And I became obsessed with uncovering what exactly they were doing differently, and why it was working. Two years and more than 130 interviews later, this book is the product of that journey.

What you're about to read is the formula that's allowed the tech giants to achieve their dominance and sustain it. This is a book about culture and leadership, but more broadly it's about ideas and invention, and the path between. It's about a new model for business in a time when companies can spin up new products in the blink of an eye, when challenge is constant and no advantage is safe. Leaning on an array of internal technology that's enabled them to operate differently, much of which they've built themselves, the tech giants have uncovered this new formula early. The time has come to release it to everyone.

The companies detailed in this book aren't perfect—far from it. In their unrestrained quest for growth, they've worked their people to the bone, missed obvious abuses of their technology, and retaliated against earnest internal dissent. Such excesses have caused the US government to consider regulation, and politicians to call for their breakup. Largely with cause. So to state it for the record: this book is not about growth, or growth hacking, or beating down smaller companies. It's about creating inventive cultures, which I believe everyone can learn from. And for those looking to rein these companies in, understanding how their internal systems work can be a strategic advantage. To effectively diagnose illness, it's helpful not only to look at the symptoms, but to understand the physiology.

If the tech giants' knowledge remained only in their hands, the broader business world—and the regulators examining them—would be at a disadvantage. In our hands, we have a chance to even the playing field.

INTRODUCTION

ALWAYS DAY ONE

At an Amazon all hands in March 2017, a trim, confident Jeff Bezos stood before thousands of his employees, looked down at a stack of note cards, and read a pre-submitted query with an expression of mild disappointment. "Okay, I think this is a very important question," Bezos said. "What does Day Two look like?"

For the past twenty-five years, Bezos had urged his employees to work each day as if it were Amazon's first. Now, with Amazon marching toward a trillion-dollar valuation, and growing by approximately a hundred thousand employees a year, a (perhaps hopeful) employee was asking Bezos to imagine Day Two.

"What does Day Two look like?" Bezos asked. "Day Two is stasis, followed by irrelevance, followed by excruciating, painful decline, followed by death."

The meeting erupted in laughter. For the thousands of Amazonians in attendance, Bezos's demolition of their unnamed colleague, who had ventured onto Amazon's third rail, was a delight. As the crowd applauded, Bezos paused, broke into a half smile, and closed the meeting. "And that is why it is always Day One," he said.

"Day One" is everywhere at Amazon. It's the name of a key building, it's the title of the company's blog, and it's a recurring theme in Bezos's annual letter to shareholders. And though it's tempting to read it as an order to work ceaselessly, particularly at the notoriously hard-charging Amazon, its meaning runs deeper.

"Day One" at Amazon is code for inventing like a startup, with little regard for legacy. It's an acknowledgment that competitors today can create new products at record speeds—thanks to advances in artificial intelligence and cloud computing especially—so you might as well build for the future, even at the present's expense. It's a departure from how corporate giants like GM and Exxon once ruled our economy: by developing core advantages, hunkering down, and defending them at all costs. Getting fat on existing businesses is no longer an option. In the 1920s, the average life expectancy of a Fortune 500 company was sixty-seven years. By 2015, it was fifteen. What does Day Two look like? It looks a lot like death.

From its origins as an online bookseller, Amazon has lived its Day One mantra, inventing new businesses with abandon, with a near-complete disregard for how they might challenge its existing revenue streams. The company remains a bookseller, but it's also a clearinghouse for almost every imaginable product, a thriving third-party marketplace, a world-class fulfillment operation, an Academy Award–winning movie studio, a grocer, a cloud services provider, a voice-computing operating system, a hardware manufacturer, and a robotics company. After each successful invention, Amazon returns to Day One and figures out what's next.

"I own a huge amount of Amazon stock," Mark Cuban told me in July 2019. "Depending on what it did today, it could literally be a billion dollars' worth of Amazon stock. And I own the stock because I see them as the world's greatest startup."

Look around at the tech giants today, and you'll see a similar path. Google started as a search website, but then invented a browser extension (Stay Tuned), a browser (Chrome), and a voice assistant (Google Assistant), and incubated a leading mobile operating system (Android). Each new Google product challenged the existing set. But by repeatedly returning to Day One, Google has remained on top.

Facebook has gone back to Day One multiple times. Starting as an online directory, the company reinvented itself with the News Feed, and it's reinventing today by moving from broadcast sharing to intimate sharing: giving the News Feed over to Facebook Groups—a series of smaller networks—and treating messaging as a first-class citizen. In the most fickle of all industries, social media, Facebook still leads.

Until recently, it seemed like Microsoft's inventing days were over. The company was so attached to Windows it almost let the future pass it by. But with a leadership change from Steve Ballmer to Satya Nadella, the company returned to Day One and embraced cloud computing, a threat to desktop operating systems like Windows, and became the world's most valuable company once again.

Apple under Steve Jobs developed the iPhone, a device that rendered desktop computers like the Mac and portable music players like the iPod less relevant but also set the company up for years of success. Today, Apple is having its Windows moment. It must leave iPhone orthodoxy behind and reinvent itself again to compete in the age of voice computing.

At Amazon's South Lake Union campus in Seattle, one of the newest buildings is called Reinvent. It's an odd word coming from one of the earth's most successful companies. But in today's business world, where Day Two is death, it's the key to survival.

Ideas vs. Execution

Operating an inventive company takes more than speeches and internal messaging. It takes a reimagination of the way you run a business, which is finally possible due to a revolution in the way we work.

There are two types of work: idea work and execution work.

> **Idea work** is everything that goes into creating something new: dreaming up new things, figuring out how you're going to make them, and going out and creating.
>
> **Execution work** is everything that goes into supporting those things once they're live: ordering products, inputting data, closing the books, maintenance.

In the industrial economy, almost all work was execution work. A company founder would come up with an idea (Let's make widgets!)

and then hire employees for execution purposes only (they'd be in the factory, making the widgets). Then, in the late 1930s, we started moving from an economy dominated by factories to one dominated by ideas—what we call the "knowledge economy."

In today's knowledge economy, ideas matter, but we still mostly spend our time on execution work. We develop a new product or service, and then spend our time supporting it instead of coming up with something else. If you sell dresses, for instance, supporting each design requires loads of execution work: pricing, sourcing, inventory management, sales, marketing, shipping, and returns. Additional support work props up these processes, including basic tasks in human resources, contracts, and accounting.

The burden of execution work has made it nearly impossible for companies with one core business to develop and support another (Clayton Christensen calls this the "innovator's dilemma"). Those who've tried have almost always pulled back, or found it impossible to sustain multiple businesses at once. "GM historically made many things other than cars," Professor Ned Hill, an economist at Ohio State University, told me, citing refrigerators and locomotives. "They were an octopus, and they couldn't manage it."

Drowning in execution work, today's companies thus devote themselves to refinement, not invention. Their leaders might desire to run inventive cultures, but they don't have the bandwidth. So they deliver a limited set of ideas from the top, and everyone else executes and polishes.

But running a company with an inventive culture, rather than one of refinement, is now suddenly possible. Advances in artificial intelligence, cloud computing, and collaboration technology have made it feasible to support existing businesses with much less execution work, helping companies turn new, inventive ideas into reality—and sustain them. These tools are the next evolution of a workplace software explosion that's made companies more efficient, and AI is putting it into overdrive. Experts say AI will free people up to do more "creative" or more "human" work. But more precisely, AI enables them to do more inventive work. This, I believe, is a critical factor behind the tech giants' success.

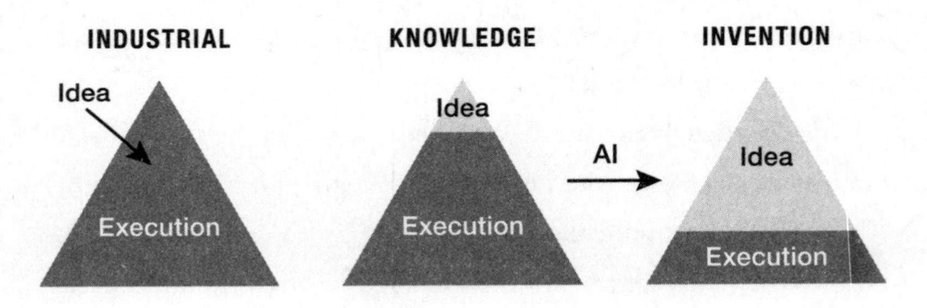

Pushing a new wave of enabling technology forward, the tech giants have figured out how to minimize execution work. This creates room for new ideas, and they turn those ideas into reality. Their cultures therefore support invention, not refinement. They remove barriers that would prevent ideas from moving through the company, and bring the best of those ideas to life. Simple in theory, but complex in practice, this is what makes them tick.

For some time, I was convinced the tech giants would hold this advantage over the rest of us for many years. But then I took a trip to Miami.

Miracles in Miami

Cee Lo Green probably never aspired to life as a corporate performer, but the hefty high-pitched singer embraced the role in October 2018 as he stood before eleven hundred badge-wearing professionals making small talk, checking their phones, and attempting to network inside Miami Beach's LIV nightclub.

As the badge wearers chowed down on sliced brisket, jalapeño mac and cheese, and blue crab risotto and indulged in an open bar, Green had some fun. He teased his top hit, a song by the name of "Fuck You" (or "Forget You" when it's on the radio), and talked up their achievements. "You're celebrating success in your life, right?" he said, moving about the stage in a white jumpsuit and sunglasses.

When the first few bars of "Fuck You" rang through LIV, the crowd went ballistic, and the wide-smiling Green fed off its energy. "If you know something you need to say 'fuck you' to, now is the time!" he yelled. A roar of fuck-yous rang back from those in attendance.

Green's performance at LIV would've been relatively nondescript, if not for the fact that it kicked off a conference hosted by UiPath, a little-known company whose software can watch your screen as you work and, with some labeling, automate your tasks. UiPath and its counterparts are on track to automate work across millions of jobs in the coming years, making the chorus of F-yous a bit jarring.

Months before the show, I heard rumblings that UiPath had the potential to change the nature of corporate work, in a manner that might bring the broader business world closer to the tech giants' way of working. And after investors handed the company $225 million earlier that fall, I decided to take my notebook to South Beach to see what it was all about.

UiPath, I learned, makes it simple to automate routine work performed on a computer. Its software can observe your mouse movements and clicks and, with some guidance, figure out how to perform your tasks. UiPath's "robots" (which have no physical presence) can take on a seemingly unlimited array of execution work, including entering data, generating reports, filling out forms, composing formulaic documents, and emailing those documents to designated recipients. In human resources alone, these bots can write standard new-hire letters, register new employees in various benefits systems, and, when the time comes, write their termination letters too.

This type of execution work fills significant portions of millions of people's workdays, and some of the world's most recognizable employers—Walmart, Toyota, Wells Fargo, UnitedHealthcare, and Merck among them—made their way to Miami to compare notes on how they might automate it.

SMBC, a Japanese bank, said it had already deployed one thousand UiPath bots and planned to add a thousand more within the year. Anoop Prasanna, Walmart's head of intelligent automation, praised UiPath's ability to automate work, complaining only that he couldn't roll out the technology fast enough. Holly Uhl, who runs automation efforts at State Auto, an insurance company, told me in a quiet moment that UiPath saved her company thirty-five thousand hours of previously human work over seventeen months, and would ramp up even further. "It's going to continue to grow," she said.

The biggest news at the conference was that UiPath's process automation technology was going to more deeply integrate with machine learning—a form of AI that can make various future-looking decisions—leading to some jaw-dropping results. Naresh Venkat, Google's head of machine learning and AI partnerships, showed off the possibilities, demonstrating how Google's machine learning could combine with UiPath's automation tech to process an insurance claim with no discernible human involvement.

In a video Venkat played onstage, a person uploaded photos of their damaged vehicle to an insurance company's website, and Google's machine-learning system reviewed the photos and determined how much the repair would cost. UiPath then opened a customer file in Salesforce, created an issue report that noted the insurance award, wrote up a basic assessment document in Microsoft Word, and emailed the assessment to the customer and an insurance company representative.

"You can automate a good chunk of what it probably takes a human to do," Venkat said, speaking with a slight hint of discomfort. "What used to take twelve days to do, in terms of processing a claim, now takes two days. It used to cost around two thousand dollars to go process something; now it costs three hundred."

UiPath is one of several "robotic process automation" companies currently surging to meet a growing demand for these capabilities. Less than two months after its Miami confab, one of UiPath's main competitors, Automation Anywhere, raised $300 million from Softbank. And Google, for its part, isn't the only company licensing AI decision-making power. A slew of other companies including Microsoft, IBM, DataRobot, and Element.ai offer similar capabilities.

With such a broad, well-funded push to bring this technology to the masses (and with demand for such technology apparent), automation likely will soon reach workplaces across the globe, taking on execution work en masse.

"You've lowered the cost of decision making—with machine learning, it will reach near zero," Forrester analyst Craig Le Clair, who studies automation, told me. "You end up with a very different workplace."

As for what this different workplace will look like, the Walmarts and Wells Fargos gathered in Miami didn't seem to know. They were eager to get automation and artificial intelligence into their workplaces, but having only dipped their toe in the water, they were left in the same place many of us inhabit today: aware that a wave of AI is coming, but unsure exactly how it will change our jobs, our companies, and the economy.

There are, however, a few companies for whom this "workplace of the future" is already a reality, and the way they're adjusting can help us understand where we're heading.

The Engineer's Mindset

The technology on display in Miami is standard inside the tech giants, and has been for years. Equipped with the world's most advanced corporate AI research divisions, these companies build machine learning not only into their products but into their workplaces as well. This technology, along with other sophisticated workplace tools, has significantly minimized their employees' execution work and increased the time they spend coming up with new ideas.

To turn these new ideas into reality, the tech giants have had to rethink the way a company is run. Loaded down with execution work, most companies today typically develop a few ideas handed down from the top, and focus on selling them. This is why "visionary" is still the ultimate compliment for a CEO today. A company's success usually rides on the ideas they and their inner circle come up with.

Bezos, Zuckerberg, Pichai, and Nadella aren't visionaries, though; they're facilitators. At the helm of Amazon, Facebook, Google, and Microsoft, they live to bring their employees' ideas to life, not their own. And they've built systems to do it. These CEOs are all engineers—not the sales or finance leaders who typically sit atop the world's leading companies—and their systems draw inspiration from their backgrounds. At the heart of their inventive cultures is something I'll call the Engineer's Mindset.

The Engineer's Mindset is a way of thinking—not a technical aptitude—that underpins a culture of building, creating, and inventing. It's based on the way an engineer typically approaches work, but it's not exclusive to any one occupation or level inside a company. The Engineer's Mindset has three main applications:

Democratic Invention

Engineers are always inventing. Their job is to build, not to sell. People employing the Engineer's Mindset appreciate that inventive ideas can come from anywhere. They build pathways to get these ideas to decision makers, and develop systems to ensure they can succeed once greenlit.

In the next chapter, we'll explore how Jeff Bezos is channeling his employees' ingenuity in a system designed to spark democratic invention and keep Amazon in Day One.

Constraint-free Hierarchy

Engineering organizations are naturally flat. Though they have hierarchy, people feel empowered to go up to the highest-ranking person and tell them precisely what they think. This is a departure from traditional organizations, where taking an idea up the chain of command is often seen as disrespecting the hierarchy.

In chapter two, we'll go inside Facebook and explore how Zuckerberg, via his feedback culture, has worked to free ideas from the constraints of hierarchy. At Facebook, employees bring ideas directly to Zuckerberg, and he processes and brings them to life. We'll also examine how his feedback system broke ahead of the 2016 election, when the company was caught off guard by election-manipulation attempts it should have anticipated, and how Zuckerberg is bringing in new "inputs" in an attempt to fix it.

Collaboration

Engineers typically work on one component of a more significant project in which, if their small thing breaks, the whole project can break down (think power grids). This type of work makes engineers master collaborators. They are regularly communicating with other groups to make sure they're working in sync. This type of mentality is well suited for bringing disparate parts of a company together to create new things.

In chapter three, we'll go inside Google and look at how Sundar Pichai is bringing people from across the company together to invent. We'll focus specifically on the collaboration it took to build the Google Assistant, which involved Google's search, hardware, Android, and AI teams, among others. The advanced collaboration tools Pichai uses to get his employees working together have also led to tribalism, trolling, and broader dissent movements within Google, which the company and its employees are still learning how to handle.

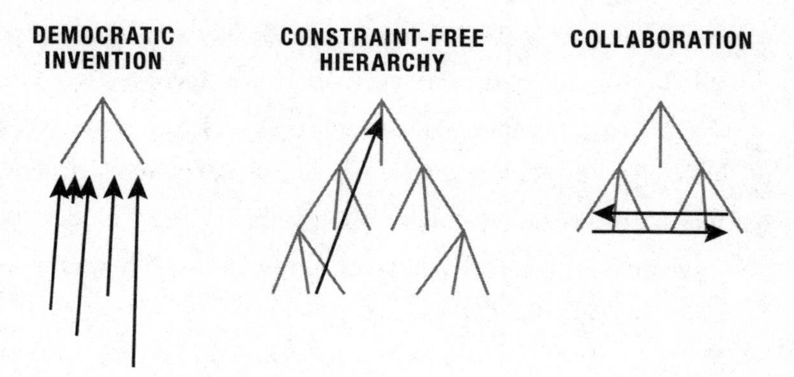

In chapter four, we'll look at Tim Cook's Apple, which is still operating in a culture built for a visionary. Apple is a company lacking democratic invention, constraint-free hierarchy, free-flowing collaboration, and useful internal technology. It's stuck in Day Two, and as iPhone sales slow, it's going to have to adjust.

We'll head to Microsoft for chapter five, where Satya Nadella is using the Engineer's Mindset to spark a new era of invention inside the company. Nadella's approach is a departure from that of his predecessor, Steve Ballmer, a case study in favor of implementing the systems outlined in this book.

The Engineer's Mindset isn't exclusively the territory of those who can code. It is, after all, a mindset, not a set of computer skills. Nor is it the province of the tech giants alone. Smaller companies can apply it just as effectively. But for now, the tech giants are ahead, especially among their tech peers. Netflix, for instance, has a feedback culture, but not one meant to spark invention. Ideas at Tesla come from the top. And Uber's culture is famously troubled.

This book will unpack the Engineer's Mindset, describing how it's the foundation of the systems Bezos, Zuckerberg, Pichai, and Nadella have built to channel ideas and bring them to life. This mindset will soon become standard in successful companies across the globe. And by reading the tech giants' stories, you'll learn how the world's top corporations are using it, providing a model you can implement in your own workplace. I hope you'll find some of the lessons worthwhile.

When Things Speed Up

As I discussed the Engineer's Mindset with the people who've lived it, the reality of today's business world began to crystallize for me. In a conversation with Sujal Patel, a trained engineer who led the data storage company Isilon Systems to a $2.25 billion exit, he laid out the following: If you're an entrepreneur today trying to bring your idea to market, all you have to do is convince one venture capitalist out of five hundred that it's a good idea, and you'll get money and build it. But if you have an idea inside a traditional company, you tell your boss, and if your boss likes your idea, they will tell their boss, and if their boss likes it, they'll tell theirs, until it gets to the top. Ultimately, if anyone along the line says no, the idea will die in the corporate muck. Meanwhile, someone off the street can bring the same concept to life.

"I always thought inside my company, 'How do I make sure that ideas that have a shot, actually get a shot?' " Patel said. "Having them flow up through a hierarchical chain is never going to work."

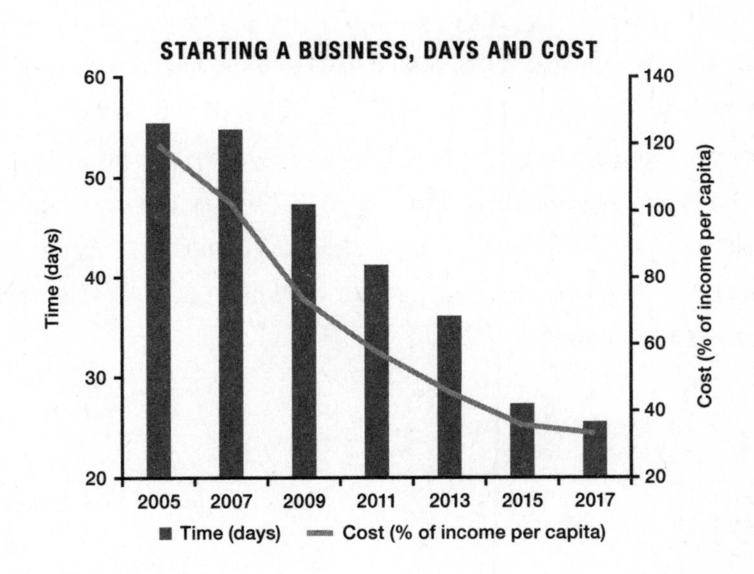

A few weeks later, the World Bank released a study that charted the cost and time required to start a new company from 2005 to 2017. In those twelve years, both factors have more than halved. When I read this, I thought back to Patel's example. If not having systems to elevate good ideas has been a liability in the past, it's an existential threat now. On one side, traditional companies are threatened by startups that can come to market quicker and cheaper than ever before. On the other, they're threatened by established companies that are operating like startups, stripping out execution work via internal technology, and bringing ideas from across their organizations to life.

This book, I believe, arrives at a transformative moment, one in which the fundamentals of work, leadership, and the overall business world are shifting. By the time you put it down, I hope you'll better understand where things are heading, and how you might want to adapt, no matter where you are on the corporate food chain. It's no secret that the CEOs in this book have been subjected to a public backlash spanning years, one rooted in discomfort and suspicion about their companies' size, power, and abuses of both. This underscores how important it is for the methods they've used to rise to prominence to be practiced responsibly. But I hope that by reading their stories, you'll find these methods less mysterious, even knowable. And if we all put them to use with care, perhaps we'll find ourselves in a more balanced economy.

CHAPTER 1

INSIDE JEFF BEZOS'S CULTURE OF INVENTION

Amazon's Seattle headquarters bear little resemblance to Silicon Valley's sprawling campuses. Rather than tuck itself away in the comfort and anonymity of the suburbs, the company operates in the heart of the still-developing South Lake Union neighborhood. Its buildings—named after project code names like Doppler (Echo) and Fiona (Kindle)—line the streets there, holding more than fifty thousand employees, with ongoing construction making room for more. Swarms of Amazonians move through the neighborhood's streets on weekdays, and if you fight your way through them, you can walk right into one of the company's most promising experiments.

A few stories beneath Jeff Bezos's office, on the ground floor of his Day One office tower, Amazon is piloting a new form of grocery store, called Go, that does away with checkout. To buy something from Go, you scan in with an app, take whatever you want, and just . . . leave. Moments later, Amazon pushes a receipt to your phone, accounting for the items you took. Go has no lines, no waiting, and no cashiers. It feels like the future, and it very well might be.

Go is powered by some impressive technology, much of which you can see by looking up. Cameras and sensors line its ceilings, pointing every which way to capture your body and its movements as you walk its aisles. Using computer vision (a subset of machine learning), Go figures out who you are, what you've taken, and what you've put back. Then it charges you. The store is almost always accurate, as I've found in my various attempts to trick it. No matter the method, be it concealing products or running in and out at my top speed (sixteen seconds total visit time), Go has never missed an item.

The story behind Go extends beyond hardware and code, though. It is, more than anything, a product of Amazon's distinct culture, the stuff you can't see. Inside Amazon, Bezos has turned invention into a habit, making the creation of new experiences like Go core to his company's business, just as critical as keeping up its famous website. Everyone at Amazon invents, from the top rungs to the bottom, and Bezos automates everything he can so they can invent more. The Amazon founder and CEO does more than encourage inventions; he's created a system meant to churn them out, giving them the best chance to succeed when they debut. Go, for instance, was initially proposed as a giant vending machine. But after going through Bezos's process, it turned into something with the capacity to change the way we shop.

Bezos's invention culture is responsible for getting us to talk to speakers, microwaves, and clocks, all with Alexa embedded inside. And to read books on screens, build companies on the cloud, shop on the internet with abandon, and, perhaps shortly, walk out of stores without stopping at a register.

"Invention is fuel for him; it's fuel for his intellect. It's part of the being, the fabric of the company," Jeff Wilke, Amazon's CEO of Worldwide Consumer and Bezos's second-in-command, told me. "The times when I've seen him most joyful are the times when he runs across an invention, an insight, an innovation, a pioneering thought."

Bezos drives Amazon's inventive culture through fourteen leadership principles, adhered to by most Amazonians more closely than their own religions, which can sometimes make Amazon feel like a cult. These principles guide decision making within the company, they're hammered home during its interview process, and they come up casually in conversation between Amazonians when they're off the clock. When you work at Amazon, the leadership principles become part of your being. They make it difficult to work at any other company, which is why so many ex-Amazonians "boomerang," or come back after they've left. One ex-employee told me he's teaching these principles to his kids.

The more you study Bezos's leadership principles, the more it becomes clear that they're a manual for invention. Taken together, they inspire new ideas, strip out the corporate muck that so often holds the best ideas back, and ensure anything with a chance to succeed gets out the door.

Think Big, for instance, encourages Amazonians to dream up the company's next great product, process, or service. And critically, it gives them permission to do it too, a departure from stay-in-your-lane management. "Thinking small is a self-fulfilling prophecy," the leadership principle says. "Leaders create and communicate a bold direction that inspires results. They think differently and look around corners for ways to serve customers."

Invent and Simplify, another example, makes invention core to people's jobs at Amazon, not peripheral. "Leaders expect and require innovation and invention," it instructs. "They are externally aware, look for new ideas from everywhere, and are not limited by 'not invented here.' "

(A more honest reading of this principle would be: Your entire purpose at Amazon is to invent. If you're not inventing, your job will get simplified and then automated. At Amazon, you invent or hit the road.)

Bias for Action tells Amazonians to get the damn thing out the door, discouraging long, drawn-out development processes in favor of producing new things. "Many decisions and actions are reversible and do not need extensive study," it says. "We value calculated risk taking."

(One Amazonian, looking for extra room in his work space, brought a saw into work and took off a chunk of his desk. When management spoke with him, he cited Bias for Action.)

Have Backbone; Disagree and Commit discourages bottlenecks by telling Amazonians to state their objection and then get out of the way. "Leaders are obligated to respectfully challenge decisions when they disagree, even when doing so is uncomfortable or exhausting," it says. "Once a decision is determined, they commit wholly."

(Bezos hated the idea of putting customer questions and answers on product pages, one ex-employee recalled, but he told the team to go ahead. Now these questions and answers are an Amazon staple.)

Finally, **Customer Obsession** puts customers before everything. "Leaders start with the customer and work backwards," it says. "Although leaders pay attention to competitors, they obsess over customers."

(Amazon's customer obsession has factored into the company's pursuit of sweetheart economic deals, its anticompetitive behavior, and its mistreatment of its employees. These activities help reduce prices and improve service, both of which often come with an unseen cost.)

If an invention isn't good enough for Amazon's customers, it gets sent back to the drawing board. "The magic of the Go store comes from the fact that once you're in, you can just walk out," one person who worked on Go told me. "[The vending machine idea] didn't eliminate the problem of checkout; it simply kicked the problem down the line." And so it was rejected.

Bezos is onto something. Inventing in today's technology-driven economy is a must, not simply nice to have. In a world driven by code, where the cost to create is lower than ever, competitors can copy what you're already doing with relative ease. To survive, you need to be creating the next big thing constantly. And so Bezos has enlisted everyone at Amazon in this pursuit. "There's invention in finance, and legal, and human resources, and fulfillment, and customer service, and every aspect of the company," Wilke said. "It becomes part of the way everybody in the company works."

Inside Amazon, Bezos has developed a culture that empowers employees to invent and lets them run the thing they've created (another leadership principle: **Ownership**). The deeper you dig in, the more apparent it becomes that this culture, bolstered by Wall Street investors who don't demand Amazon turn a profit, is what's behind the company's array of beloved products and services: Echo, Kindle, Prime, Amazon Web Services, and Amazon.com. It is, in no uncertain terms, Amazon's competitive advantage.

Meet Amazon's Science Fiction Writers

On June 9, 2004, at 6:02 P.M., Jeff Bezos banned PowerPoint at Amazon.

Not one for subtlety, he delivered the news right in an email subject line: "No powerpoint presentations from now on," he wrote to his senior leadership team. PowerPoint, Bezos understood, is a terrific selling tool, making mediocre ideas look great by dressing them up in bullet points and fancy templates. For the same reasons, it's terrible for inventing, giving people "permission to gloss over ideas," as he put it, often yielding flawed or incomplete concepts, even if it didn't seem so at the time of presentation.

Bezos offered an alternative: written memos. Instead of slideshows, he wanted Amazonians to write up ideas for new products and services in documents composed of paragraphs and complete sentences—no bullet points allowed. These memos would be comprehensive, making it easy to spot gaps in thinking, and they'd help Amazonians' imaginations run wild as they composed them. "The narrative structure of a good memo forces better thought and better understanding of what's more important than what, and how things are related," Bezos wrote.

It's one thing to have values, and the leadership principles are a clear articulation of Amazon's, but without a system through which employees can put those values into practice, they're often worth little. The moment Bezos hit Send on that email, he laid the foundation for Amazon's system of invention, one that puts the written memo at the center.

Today, all new projects inside Amazon kick off with memos. Set in the future, these memos describe exactly what a potential product will look like before anyone starts working on it. Amazonians call this "working backwards." They dream up the invention first and then work backward from there. Limited to six pages, the memos are typically single-spaced, typed in eleven-point Calibri font, a half inch at the margins, and picture-free, and detail everything you could want to know about a proposed new product and service.

I got a chance to look at one of these "six-pagers" when I was in Seattle, given access by an ex-Amazonian who asked to remain anonymous because they were supposed to have deleted it. The memo was exhaustive, containing an overview of the proposed new service, what rolling it out would mean for customers, what it would mean for Amazon's vendors, a financial plan, an international plan, pricing, a work schedule, revenue projections, and metrics for success.

Writing these memos is like writing science fiction, one ex-Amazonian told me. "It's a story, set in the future, of what you believe the future is going to be," he said. "It's a story for something that doesn't exist." There's actual fiction involved too: The six-pagers often contain fake press releases announcing the would-be product to the world, complete with fake quotes from executives hailing its arrival.

When an Amazonian's six-pager is ready to be reviewed, they ask for a meeting with the senior leaders who can help them turn their science fiction into reality, and then things get a little weird. With no PowerPoint to talk through, meetings inside Amazon kick off with silence. For fifteen minutes to an hour, everyone in the room quietly reads through the memo, takes notes, and prepares to ask questions. It's agony for the memo writer, who has to sit there and watch Amazon's top leaders, sometimes Bezos himself, comb through their ideas without making a peep. "I don't get thirty minutes with Jeff every week or every month," Sandi Lin, a former Amazon senior manager, told me. "I get one shot to present my ideas."

"You put months of work into it," Neil Ackerman, an ex–Amazon general manager who's written multiple six-pagers (and has eight patents to show for it), explained. "For the first hour you sit down, you give everyone the pieces of paper, stapled, with a highlighter and a pencil—you don't mail it out beforehand because no one pre-reads; that's bullshit—then they basically are quiet for a whole hour and everyone reads," he said.

After the reading period, the most senior person in the room opens the floor up for questions, then the people sitting around the table show no mercy. "Now they sit there for another hour," Ackerman said. "Then they get question and answer, question and answer. They get constantly barraged with questions—if it gets approved, they have a project."

When a six-pager is approved, Amazon gives the person who wrote it a budget to start recruiting and build the invention they've dreamed up. Putting the person who wrote the six-pager in charge of bringing the idea to life is critical to Amazon's ability to invent,

Micah Baldwin, an ex-Amazonian who's been through the process, told me.

"There are two sides to invention," he said. "There's thinking and doing. And most doers don't think. And most thinkers don't do. And the great thing about a narrative is it forces you to do both. I have to think through the idea beginning to end—who cares, who wants it, who's the customer, the whole deal—and I need to be able to put it in a narrative format that if I give it to you clean, you've never seen this ever, you can have an opinion about it, and be supportive of it or not. And then it's my responsibility to execute it once it's done. I'm not just writing it as a think piece. I'm forced to think, and I'm forced to do. Those two things in combination drive innovation."

The six-pager democratizes invention within Amazon. Anyone inside the company can write one, and if they build enough traction, senior leadership will review it. "I read six-pagers that come from other parts of the company that don't report to me," Wilke told me. "I read six-pagers that are from people who are multiple levels down in the traditional organizational hierarchy. They can come from anywhere."

The deep detail in these memos makes it easy for Bezos and his lieutenants to understand a project, approve it, reject it, or send it back to the team for further development. In this system, Amazon's employees drive its success; they're constantly improving, tweaking, and inventing through six-pagers, with Bezos acting as facilitator.

It may seem strange, or even a stretch, to describe a corporate culture as one of invention. Employees typically focus their energy on keeping the company running, not writing science fiction. Amazonians have vendor relationships to manage, warehouses to stock, and products to ship. How can they focus on inventing? Well, that's where the robots come in.

Jeff Bezos's Robot Employees

A few thousand miles east of Amazon's downtown Seattle headquarters, a large beige-and-gray warehouse sits not far off the highway. The building is the kind of size you measure in football fields. For the record, it can hold about fifteen, more than an entire Sunday's worth of NFL games.

Named EWR9 after the nearby Newark Airport, the warehouse is one of more than 175 fulfillment centers, or FCs, that Amazon uses to store, pack, and ship millions of products to customers every day. EWR9 itself can ship hundreds of thousands of packages in one twenty-hour operating period.

When I visited EWR9 on a hot summer day in August 2018, the FC hummed with the sounds of robots working alongside human "associates." The robots—small orange Roomba-like machines—moved about the cavernous FC with purpose. They slid underneath tall yellow racks of products, lifted them, rotated, and shuttled them between storage and human workers. They moved with such coordination it seemed as if they were dancing.

The robots are the most visual example of Bezos's obsession with automating whatever he can to free his employees to work on more creative tasks. "I don't think I can remember a time where he wasn't interested in using computing to help us achieve our mission," Wilke said. "From the earliest days, he would look at a process, and if there was repetitive work being done by people who can be freed up to be more inventive, he would say, 'How do we automate that process? How do we automate that routine so that our people can be as creative as possible?' "

When I walked into the fulfillment center, I was greeted by Preet Virdi, EWR9's charismatic general manager (who's since moved to the same role in Baltimore). Virdi is tall and exceedingly friendly, has a booming voice, and is a true company man. He's the type of guy you could imagine crying tears of joy as he learned about Amazon's dedication to customers during new-hire orientation. Over the course of my visit, Virdi delivered a steady stream of optimistic observations without a hint of irony. "Anytime you get a chance to work with people, it's just an awesome experience," he told me. "Amazon robots work with Amazon associates in a really cool and nice manner." Amazon is a publicity-averse company, but it's clear why it lets Virdi out into the wild.

Aside from being a walking press release, Virdi is a new type of manager, leading humans who work alongside robots, a dynamic Amazon has spent the past eight years figuring out how to navigate. In March 2012, Amazon acquired Kiva Systems, the manufacturer of the robots used in its FCs, and has since deployed the bots with extraordinary speed. Amazon placed roughly 15,000 robots in its FCs by 2014 and had 30,000 in operation by 2015. Today, the company employs more than 200,000 robots, augmenting a workforce of approximately 800,000 humans. At EWR9, approximately 2,000 people work alongside hundreds of robots.

Robots have changed the way FCs operate. Before their introduction, FCs relied on humans to canvass Amazon's giant warehouses, find products its customers had purchased, and walk them back for shipping (in those Amazon FCs without robots, this is still how it works). Robots handle that work today, and as robotics technology advances, Amazon seems likely to automate other core parts of FC work. For now, human "stowers," "pickers," and "packers" remain. Stowers load up the racks, pickers grab items people have purchased off them, and packers pack those items into the boxes and envelopes that show up at your door. In between these activities, the robots drive off to the "robotics floor," depositing their racks beside thousands of others waiting for their next ride aboard a Roomba on steroids.

The cooperation between human and robot is something to behold. When you order a product on Amazon, a robot drives over to the rack containing it, slides underneath, picks up the rack, calmly lines up with other robots next to a workstation, shuffles in front of a human worker when Amazon's software tells it to, remains in place until the worker grabs the product, and then scoots away. As I watched a picker work, the efficiency struck me. He grabbed a product from a rack, dropped it into a bin, the robot scurried off, the next robot stopped by, a section of the rack lit up, he grabbed a product from that section, and off the robot went. It all moved very quickly.

Advanced software under the hood makes the process run smoothly. The robots move through the FC by reading QR codes scattered across the floor. When a robot passes over a code, it's instructed either to wait or to move to the next QR code, where it's given more instructions. The system knows how fast each picker and stower works, and automatically sends more robots to the faster workers and fewer to the slower ones. At another FC I visited, in

Kent, Washington, the robots stop in front of cameras that scan the racks, assess the amount of space left (using computer vision), and determine when they should be sent back for more stowing (or sent to a problem-solving team when items look askew). As pickers work, some compete voluntarily in "FC games," which rank them on speed.

The employees I met at the two FCs I visited seemed to be in good spirits and happy to work at Amazon. But that's not the case everywhere. James Bloodworth, a British journalist who went undercover at an Amazon FC while conducting research for his 2018 book, *Hired: Six Months Undercover in Low-Wage Britain*, said he once found a bottle of urine on the floor, apparently left there by an associate so afraid they'd miss productivity targets they didn't want to take a bathroom break.

Amazon works its people hard. Things get especially intense ahead of its Thanksgiving and Christmas peak season. The company's corporate staff even used to take shifts inside fulfillment centers to help meet the demand. Robots should theoretically take some of the slack off Amazon's workers. But they also make you wonder if being an out-of-work associate is preferable to being an overworked associate.

Inside the Kent FC, a picker named Melissa, tattooed, in her twenties, and also a part-time Starbucks employee, told me she anticipates Amazon will one day automate more work in its fulfillment centers. "There's going to be a way here where you won't need somebody filling your totes all the time," she said, using Amazon's term for bins. When I began bringing this up with Preet Virdi, it resulted in a fairly awkward exchange.

"Amazon doesn't like repetitive work," I said. "It's pretty clear having studied the company a little bit—"

"Not sure what you mean by that," Virdi said, cutting me off. "Can you explain repetitive work?"

"If things are repetitive and low value add—"

"Okay . . ."

"You don't like the premise of the question."

"I just want to understand repetitive work," Virdi replied. "When we pick orders, that could be considered repetitive work. So is packing boxes and shipping boxes. That's repetitive work, but that's why we are here."

I asked Virdi what Amazon associates could do if their work was automated. He said there were two options: They could find another, similar skilled job inside an FC, like packing boxes. Or they could take a training course to learn how to do something more technical. With three to four weeks of training, he said, they could get a job within Amazon as a robotics floor technician. "That job did not exist in traditional fulfillment centers," Virdi said.

The more time I spent in EWR9, the more jobs I heard about that never existed before. There are robotics floor technicians, amnesty professionals (who clean up after robots when they drop products), ICQA members (who count the items in the racks, making sure they align with the system's numbers), and quarterbacks, who monitor the robotics floor from above. In the same time Amazon has added the two hundred thousand robots, it's added three hundred thousand human jobs.

Amazon's push toward automation may not be sending its

associates to the unemployment lines, but it is forcing them to navigate constant change, which can be both invigorating and exhausting. When you work at Amazon, you could be doing something one day, only to have it replaced by computers or robots the next. "You have to coach and teach people how to be lifelong learners," Wilke told me. "The way that you reward work, and learning, and how much of people's time is devoted to these things is changing."

Amazon walks the walk in this regard. It lets its employees know what's coming and offers training to help them take on new jobs. An Amazon training course called A2Tech, for instance, teaches associates how to perform technical work inside the FCs through instruction, hands-on work, and exams. Career Choice, another program, will pay 95 percent of an FC worker's tuition for classes counting toward certain degrees and certificates, capped at four years or $12,500.

The constant change can be hard for people who aren't adept at handling it, as Bezos has acknowledged. "Because of the challenges that we have chosen for ourselves, we get to work in the future. And it's super fun to work in the future, for the right kind of person," he said in a 2016 interview with the veteran tech journalist Walt Mossberg. "For somebody who hated change, I imagine high tech would be a pretty bad career, it would be very tough, and there are much more stable industries, and so they should probably choose one of those more stable industries with less change, and they'd probably be happy there."

It's a nice idea, but the people moving into those professions shouldn't get too comfortable either, as the type of change Amazon workers are experiencing is on track to be felt everywhere. Even the insurance claims adjuster, an example Bezos gave of a steadier profession, is subject to automation, a conclusion he reached almost as soon as he offered the example. When Mossberg told Bezos insurance claims adjusters have iPads now, Bezos replied, "Pretty soon they'll have machine learning too." He was right. Insurance companies are already using machine learning to calculate home insurance rates and to monitor driver safety. And, as my visit to UiPath's Miami confab illustrated, these systems seem destined to replace the insurance adjuster altogether.

Amid the change, one thing remains constant at Amazon: the determination to invent. Automating so much labor has freed Amazon's corporate staff to concentrate on its invention process (they no longer have to work peak season, packing boxes). And it gives Amazonians inside the FCs the time to invent on their own. In EWR9, Virdi showed me a "continuous improvement" kiosk where employees enter ideas for new products, processes, and minor FC tweaks. Every Wednesday, Virdi and his senior staff review the best ideas for forty-five minutes. When they like what they see, they give their associates time and resources on their scheduled workdays—paid—and ask them to turn their ideas into reality. Associate feedback, to give one small example, led Amazon to make its bins yellow to help them more easily spot products and work more effectively.

As our walk through EWR9 wrapped up, I bought lunch for Virdi, myself, and an accompanying Amazon spokesperson from a Chick-fil-A stand near the facility's entrance. I had the spicy chicken sandwich, as did the spokesperson. Virdi took the plain one. The robots kept working.

Hands off the Wheel

Humans are incredibly predictable, a fact that Amazonians know well. "Pick a ZIP code, and Amazon can pretty much tell you what people wear, buy, and do in that ZIP code," Neil Ackerman, the ex–Amazon general manager, told me. "Go from house to house. They wear the same clothes, they eat the same food, they decorate the same, they buy the same stuff. It could be different colors, but they're mostly predictable."

With twenty-five years of historical data at its disposal, Amazon knows what we want, when we want it, and it's likely already sent the next thing you're going to order to a fulfillment center near you, ready to ship when you hit Buy. Amazon knows a surge in winter coat orders is coming in the fall. But not only that, it knows certain ZIP codes buy lots of North Face jackets, so it can load up on North Faces in the nearby FCs.

Using this knowledge, Amazon is automating a wide array of work in its corporate offices under an initiative it calls Hands off the Wheel.

Amazon's fulfillment centers get stocked with products before people buy them, a necessity for a company offering two-day shipping to more than 150 million paying Prime subscribers (with one-day shipping now rolling out). Traditionally, Amazon has employed "vendor managers" to make this process run smoothly. A vendor manager working with Tide, for instance, figures out how much detergent to place at each Amazon fulfillment center, when it needs to be there, and how much Amazon will pay per unit. Then, they negotiate the price with Tide and place the orders. The position was prestigious inside Amazon until recently. It was fun, relationship-oriented, and put Amazonians in contact with the globe's top brands. But at Amazon, change is always lurking around the corner.

In 2012, Amazon's senior leadership began examining whether some of the core tasks performed by vendor managers needed to be done by people. If the humans were predictable, Amazon's algorithms could potentially determine what products needed to be in which fulfillment centers, when they needed to be there, in what quantity, and for what price. And they could probably do it better than humans.

"Buyers in a traditional manner do the same thing over and over again," Ackerman said. "They get a call, they get a sales pitch, they buy an amount of product, they usually buy the wrong amount because they're humans, and lo and behold, people buy the products, and it's a cycle. When you have actions that can be predicted over and over again, you don't need people doing that. And frankly, computers, or algorithms, or machine learning, are smarter than people."

Understanding this, Amazon's leaders decided to attempt to automate traditional vendor manager responsibilities including forecasting, pricing, and purchasing. People inside Amazon began calling this initiative Project Yoda. Instead of having vendor managers do the work, Amazon would use the Force.

In November 2012, Ralf Herbrich joined Amazon as its director of machine learning. One of his early goals was to get this project off the ground. "I remember when I started. We had a lot of manual decision making and forecasting still happening," Herbrich, who left Amazon in late 2019, told me via phone from Berlin. "We were starting to look into algorithms; in fact, it was one of my launch projects."

Herbrich and his team—which fluctuated from a few dozen machine-learning scientists to more than a hundred—spent the next few years hunkered down trying to bring Project Yoda to life. They initially tried a few textbook machine-learning approaches, which were ideal for predicting orders for products purchased at high volume, but broke when applied to products purchased sporadically. "They worked nice on maybe one hundred or one thousand products," Herbrich said of the textbook methods. "But we had twenty million to do." So they tinkered. Each time Herbrich's team came up with a new formula, they used it to simulate the previous year's orders, trying to figure out how it did versus Amazon's actual numbers with human workers.

With trial and error, Herbrich's predictions got good enough that Amazon began putting them right in its employees' workflow tools. Vendor managers now saw predictions for how many units of products to stock in each region. Vendor managers (and their colleagues who assisted with product ordering) used these systems to "augment their decision making," as Herbrich put it.

In 2015, the initiative once known as Project Yoda became Hands off the Wheel, whose name says almost everything you need to know about its intentions. Instead of simply taking the machine-learning algorithms' predictions into account as they made their decisions, Amazon vendor managers were instructed to take their hands off the wheel and let the system do its work. Soon, Amazon's senior leadership team set high goals for the percentage of retail employee actions that should be entirely hands off the wheel. Manual interventions were discouraged, and in some cases had to be approved by category managers (quasi-CEOs in their own right).

Vendor managers' jobs soon changed profoundly. "We were not able to order as much as we used to with the freedom and flexibility that we used to be able to," Elaine Kwon, a former vendor manager, told me. "At some point, if I'm preparing for a big holiday, I spend a lot of my time thinking about what to order. That is a buyer's job, to figure out what to buy. That's slowly getting taken away too. [Management] was like, 'No we're not going to have you guys do that.' "

At a meeting about Hands off the Wheel goals, an ex–Amazon employee I'll call Tim, who asked to remain anonymous, fearing retribution, said he brought up what was quickly becoming evident to him. "So, just to be blunt, we should probably find different job functions, because we're clearly working ourselves out of a job here?" he said. The room laughed, but Tim was serious. And eventually, the presenter said yes, this would decrease the amount of human involvement. "They essentially said yes but didn't want to put too fine a point on it," Tim explained.

Hands off the Wheel eventually extended across the entire retail organization. Forecasting, pricing, purchasing, and inventory planning are now being done with the assistance of, or entirely through, automation. Merchandising, marketing, and even negotiation are also partially automated inside Amazon. When suppliers want to make a deal with Amazon, they now often negotiate with a computer portal instead of a vendor manager. The car is driving itself.

Life After Yoda

Stories like this usually go the gloom-and-doom route from here, diving into mass unemployment, the end of work, and eventually end-time. We may get there. But as I spoke with Amazonians who've experienced Hands off the Wheel, I was surprised by their matter-of-fact attitude and lack of concern about what this portends.

"When we heard ordering was going to be automated by algorithms, on the one hand, it's like, 'Okay, what's happening to my job?' That's definitely a question you're thinking about," Kwon said. "On the other hand you're also not surprised, you're like, 'Okay, as a business this makes sense, and this is in line with what a tech company should be trying to do.'"

Tim, a bit saltier, netted out in the same place. "It was a total change," he said. "Something that you were incentivized to do, now you're being disincentivized to do it. . . . It's a little heartbreaking, you know, working yourself out of a job. But it's difficult to disagree with the logic."

Another current employee told me that at Amazon, "you're constantly trying to work yourself out of a job. You should not be doing the same thing day to day. Once you've done something consistently, you need to find mechanisms to invent and simplify."

From a business standpoint, it's easy to see why Amazonians (who receive a chunk of their compensation in stock) feel this way. Amazon's business is a flywheel—a self-reinforcing system that gets better and stronger as each component improves. By offering a wide selection of products at low prices, and a convenient shopping experience,

Amazon generates traffic from people looking to buy products. The traffic makes Amazon a more enticing place for sellers, who sell more products at better prices to reach Amazon's customers, generating more demand. And so the flywheel spins.

In Amazon's early days, with fewer seller relationships to manage, the company could hire humans to manage its vendor relationships. But as Amazon scaled to twenty million products, the labor cost to manage every relationship with a human would be prohibitive, causing prices to rise and sticking a wrench in the flywheel.

"You could start the Amazon business with doing it in an untechnical way. But you couldn't scale it," Herbrich said. "Each of those processes in our flywheel, pretty much each of them, will really only scale up if we are automating some of those decisions that people are doing, particularly the ones that are based on repeated patterns that we observe, and that's where AI comes in."

Twenty years ago, an Amazon vendor manager could handle a few hundred products, Herbrich said. Today, he said, they're working with anywhere from ten thousand to one hundred thousand. (An Amazon spokesperson said Herbrich was using these numbers as an example and they shouldn't be taken at face value.)

When Amazon automated its retail employees' forecasting, purchasing, and negotiation duties, it didn't eliminate their jobs, but it did fundamentally change them. Vendor managers are now more auditors than doers. "They go from typing to selecting," Herbrich said. "When there are mistakes, often what we find is that they now need to have the skills to diagnose what inputs to the algorithm may be wrong. It shifts from making the outputs, how many units to buy, to changing the inputs."

If that sounds like gibberish, here's an example of how it plays out in real life: Amazon's inventory forecasting system was once missing predictions on some basic fashion products. Herbrich was incredulous; white socks should not be something that's hard to forecast. So he ordered a review of the inputs going into the prediction tool, including color, and found that Amazon had fifty-eight thousand different color categories in total. Spelling mistakes and nonstandard spellings had thrown the system off, and when they standardized color, things went back to normal.

By overriding bad predictions—that is, putting their hands on the wheel—Amazon's employees would paper over problems with the inputs that drive the algorithms. Fix the inputs (in this case, normalize color categories) and you fix the system.

Hands off the Wheel is moving beyond Amazon's vendor managers and marketers. Amazon's language translators are now becoming machine-learning auditors too. Instead of translating product pages themselves, Amazon's translators are now monitoring automated translations, presented by Amazon's systems with a degree of confidence that they're accurate, and making changes as necessary. When you go to order something from an Amazon product page originally written in another language, it's often impossible to tell whether it was translated by a human or AI.

Machine-learning translations again help spin Amazon's flywheel. The more languages a seller can sell in, the more selection there will be for Amazon customers, and the more they'll visit the site. Thanks to this added traffic, sellers will be more eager to work with Amazon and offer more products at better prices, which will bring in more customers.

While much of what we do is predictable, there are some things that algorithms can't account for, such as taste. To make up for this, vendor managers are also doing some creative work in addition to auditing automated forecasts, negotiations, and orders. "In our fashion business, rather than have our team that's really good at detecting new trends sitting in an office in Seattle working on spreadsheets, I'd rather have them going to shows in New York, Milan, and Paris and detecting the very latest trends in supersophisticated ways," Wilke said. "You get the best of both. You get scalability from the computing engine, and you get insight and intuition that only humans can provide."

Fads can also make the retail industry entirely unpredictable at times, requiring humans to stay on top of them. "If you think about the real world and its products, this is not a static space," Herbrich said. "Fidget spinners didn't exist in 2016, and they exist now. Maybe they won't exist anymore in 2020 or 2021. You need to constantly find ways to capture features of new things in the world."

Thanks to Hands off the Wheel, Amazon's retail division now operates more leanly and efficiently. The concept has also enabled Amazon's third-party marketplace and fulfillment operation—where vendors list directly on Amazon, instead of relying on Amazon as a middleman—to thrive.

The prestige of the vendor manager job has worn off a bit too, and many vendor managers have moved to new roles within Amazon. When I browsed LinkedIn to find out where they went, I found many ended up in two specific job categories: program manager and product manager. Program and product managers are professional inventors at Amazon. They dream up new things and steward them along as they get built. Product managers typically focus on getting individual products built, and program managers focus on multiple, interrelated projects. According to LinkedIn's data, these are the fastest-growing job functions inside Amazon today. "That was a thing that a lot of people really looked for," Kwon told me. "They were also looking for other cool teams that valued innovation."

Tim noticed a similar migration. "I have friends in categories where two years ago there were twelve people in that category, now there's three," he told me. "Almost every person that I knew in retail at this point has a job now that is product manager, program manager. Nobody is really in a core retail function anymore. If you're a non-engineer, you become a program manager or a product manager."

By automating work inside its retail division, Amazon is opening up new opportunities for inventing, which has been the plan all along, according to Wilke. "People that were doing these mundane repeated tasks are now being freed up to do tasks that are about invention," he said. "The things that are harder for machines to do."

In 2011, Amazon's VP of pricing and promotions, Dilip Kumar, left his position in the retail organization to shadow Bezos for two years as his "technical adviser." Amazonians covet the technical adviser position. Those who inhabit it attend every meeting Bezos takes, getting a chance to look at Amazon through his eyes, and often gain license to take a major swing when they're done. Andy Jassy, Bezos's first technical adviser, went on to found Amazon Web Services (AWS), Amazon's cloud services division, which now brings in some $9 billion per quarter with Jassy as its CEO.

Kumar—whose LinkedIn profile describes his time as Bezos's technical adviser as "Quite possibly the best job that I've ever had!"—also set out to do something big upon completing his tour of duty. But while he was away from Amazon's retail division, Project Yoda began automating pricing and promotions, his domain of expertise, freeing him up (or forcing him) to try something new.

Kumar, along with some others from the retail organization, set out on a mission to find the most annoying part of shopping "in real life" and attempt to fix it using technology. They settled on checkout. And after a few tries, including a concept for a very big vending machine, the team developed Amazon Go.

Insist on the Highest Standards

You might wonder when Bezos, CEO of a trillion-dollar company and a billionaire many times over, will be ready to ease up a bit. He's built a world-beating e-commerce business, a thriving hardware division, an Oscar-winning movie studio, and a massive enterprise software business. If Bezos wanted to, he could ride these businesses for decades, generating a small-country GDP's worth of money each year, and make room for some upstart, similar to what he was twenty-five years ago, to give it a go. That is unlikely to happen any time soon.

Bezos's joy in life comes from his work, specifically the inventive parts that make him feel like he's back in the nineties, trying to figure out how to sell books on the internet. For many ultrasuccessful CEOs, living the good life means spending your days on private islands or coasting around the globe on a boat. For Bezos, the good life is work, and coasting is "excruciating, painful decline, followed by death."

There's an emotional need that fuels Amazon's top leaders, Wilke told me. "I want to invent, and be in uncharted territory, and have the emotion associated with the unknown that is a mixture of fear, and uncertainty, and excitement, and the belief that if you push through whatever barriers are in the way, that you end up in a state that is amazing," he said. "That kind of keeps you going, and I bet it keeps him going too."

It's no coincidence that the people who run Amazon, the leader in the notoriously ruthless e-commerce sector, are driven by the high they get from inventing, a force more powerful than sales quotas or Wall Street expectations. If Amazon eased up even slightly, a competitor could come in and ship faster, or offer lower prices, or provide a better shopping experience, and Amazon's flywheel would grind to a halt. Customers would migrate to its competitor's website (it's as easy as typing in a URL), and that additional traffic would win its competitor more suppliers, enabling it to reduce prices and offer a greater selection, attracting even more customers as a result—and those are Amazon's customers.

"Customers are always unsatisfied," Bezos said in an April 2018 interview. "They're always discontent, they always want more. No matter how far you get out there in front of your competitors, you're still behind your customers. They're always pulling you along."

Bezos's urgency has fueled Amazon's rise. It's also, at times, put incredible pressure on his employees, who feel the need to keep up. Another Bezos leadership principle, **Insist on the Highest Standards**, spells out the company's expectations pretty clearly: "Leaders have relentlessly high standards—many people may think these standards are unreasonably high."

One ex-Amazonian I spoke with told his wife and kids, "Daddy's going to war," when he accepted the job at Amazon. He ended up working straight through Thanksgiving dinners in service of the company's goals. Sandi Lin, the former Amazon senior manager, repeated a saying from her time at the company that captures the mentality well: "At Amazon, if you turn water into wine, the first question is, 'Well, why wasn't that champagne?' "

"That Article"

When setting such high standards, Amazon implicitly encouraged unhealthy workaholism. And on August 15, 2015, the worst elements of Amazon's culture spilled out into the public in a brutal five-thousand-word *New York Times* article headlined "Inside Amazon: Wrestling Big Ideas in a Bruising Workplace."

The story, still known as "That *New York Times* article" among Amazonians, painted Amazon as a ruthless, draconian place to work: Its employees were routinely subjected to searing criticism and encouraged to dress down their colleagues via custom-built feedback tools. They were forced to work long hours, straight through holidays, vacations, and weekends. They were put on "performance improvement plans" after bouts with cancer and stillbirths, told to check their personal problems or get out. They were largely miserable. Toward the top of the article, the *Times* splashed an ex-employee's quote that Amazonians still talk about today: "Nearly every person I worked with, I saw cry at their desk."

When the article came out, phones started ringing inside Amazon's

South Lake Union headquarters. "Are you okay? Are you doing all right?" Elaine Kwon, the former vendor manager, was asked by a nervous brand contact in New York. Micah Baldwin, at the doctor's office, was asked by his physician if he'd cried at his desk (he had not).

Amazon went to war with the *New York Times* following the article's publication, seeking to discredit its sources. In a Medium post titled "What *The New York Times* Didn't Tell You," Amazon SVP and former White House spokesperson Jay Carney took specific aim at the ex–Amazon employee who told the *Times* about all the crying. "His brief tenure at Amazon ended after an investigation revealed he had attempted to defraud vendors and conceal it by falsifying business records," Carney wrote. "When confronted with the evidence, he admitted it and resigned immediately."

New York Times editor Dean Baquet shot back: "His one quote in the story was consistent with those of other current and former employees. Several other people in other divisions also described people crying publicly in very similar terms," Baquet wrote in a reply, also on Medium. In a matchup between heavyweights, both sides landed punches.

When the article hit, Bezos emailed the company. "I don't recognize this Amazon and I very much hope you don't, either," he said. "Even if it's rare or isolated, our tolerance for any such lack of empathy needs to be zero." Amazon has long said it hasn't changed its systems, mechanisms, or principles based on the *New York Times* story. But after the article was published, the company got to work dealing with its problems.

Amazon expanded its use of daily surveys called Connections, meant to figure out where its culture needed improvement. The surveys asked questions like: When was the last time you had a one-on-one with your manager? Does your manager reflect X value principle? (The tool originally debuted in North American fulfillment centers in 2014.) Using the data collected from this survey, Amazon began running a version of its process on its own culture.

"Amazon is obsessed with inputs and outputs. And the output is: the *New York Times* writes an article. That's not a great output. The inputs are what got you there," one ex-employee told me. The inputs, he explained, were all the little things the survey was testing for. "You're working backwards from the output that you want, and you're inventing a new tool and a new process to do that."

Amazon then made some changes. It simplified its review process, which previously required lengthy self-evaluations from employees, structured around the leadership principles and spanning more than a dozen pages at times. Now its employees simply list their "superpowers." Amazon also simplified its promotion process, which previously required managers to fight for their reports in front of skeptical colleagues, who ultimately decided the matter. If a manager didn't want to fight for you, or recently spent their political capital fighting for a different employee, you could lose out on a promotion even if you were overperforming. Amazon has since simplified the process by allowing managers to submit someone for promotion via a software tool. Amazon also deep-sixed its **Be Vocally Self-Critical** leadership principle, moving elements of it to **Earn Trust** and adding a new principle, **Learn and Be Curious** (its Vocally Self-Critical a cappella group survives). Today, average tenure at Amazon is up from before the *New York Times* article, one current senior Amazonian told me.

Aside from fielding phone calls from concerned relatives, friends, and business partners, as well as in-person queries from doctors, the *New York Times* article was strange for some Amazonians for another reason: this is what they signed up for. Most Amazonians I spoke with joined the company expecting to work hard. They embraced the opportunities and challenges that came along with working in a company that will allow you to take on as much responsibility as you think you can handle, will let you work as hard as you're willing, and believes you're capable of bringing your ideas to life. When Amazonians leave for jobs at Google or Microsoft, their Amazon colleagues wish them well in their retirement.

"In that *New York Times* article, when you read about people crying at their desks, internally, people were like, 'They're just not Amazonians.' They weren't tough enough," one ex-employee told me. "Part of the toughness of Amazon actually is appealing to the people that work there. We don't want free lunch, you know what I mean?"

Outputs

As I spoke with Amazonians, current and former, they seemed more interested in what working under Bezos would turn them into, as opposed to the work they were doing itself. By living inside Bezos's culture of invention (with a big assist from the AI systems taking over the tasks they'd otherwise spend their time completing) they learned to be creative in a technical way.

In our society, there's a belief that technically minded people can't think creatively, and that creative people can't think technically. We have artists and musicians in one column, and coders and mathematicians in the other: the right brain versus the left brain. At Amazon, Bezos teaches people how to bring these two together. He makes them imagine the future, write the science fiction, code it up, automate, and imagine the next one. And he may kick their ass in the process.

Many of those I spoke with are using this technical creativity in their next steps in life. Sandi Lin, the former senior manager, is the founder and CEO of Skilljar, an online training company that's raised more than $20 million. Elaine Kwon is the cofounder and managing partner of Kwontified, an e-commerce software and services company, and she's now working on automating certain employee tasks, similar to how Amazon automated some of hers as a vendor manager. Micah Baldwin is helping startups in a role at Madrona Ventures, a Seattle-based venture capital firm. Neil Ackerman now manages supply chain at Johnson & Johnson, where's he's working to bring elements of Amazon's process into the 130-year-old manufacturer. Ralf Herbrich joined German e-commerce company Zalando in January 2020 and is ready to infuse it with machine learning, just like he did to Amazon. Jeff Wilke, meanwhile, is poised to take over from Bezos when he eventually steps down, though it may be a bit of a wait. And Preet Virdi is most likely walking around an Amazon FC right now, smiling as he engages its associates.

In the one opportunity I had to stroll through Amazon's headquarters, signs of what Bezos is shaping his employees into were everywhere. There were posters heralding inventions like Echo and Prime Now, autographed by the teams that brought them to life. There was a giant word search on a wall with one word highlighted at its center: *invent*. There were three giant glass structures, the Spheres, filled with rare plant species, and plenty of small work spaces, meant to inspire creativity.

On the bottom floor of Amazon's appropriately named Day One tower, the building where Bezos works, the company's first Amazon Go store is operating and open to the public. As I watched its exit, I saw a stream of bewildered tourists walking out, looking around, trying to figure out how exactly Bezos does it.

CHAPTER 2

INSIDE MARK ZUCKERBERG'S CULTURE OF FEEDBACK

On a sunny Monday morning in Menlo Park, a group of thirteen Facebook employees gathered in a large, open room for a lesson in the delicate art of calling out your coworkers.

The group—made up of managers and individual contributors, engineers and marketers—took seats, exchanged smiles, and settled in somewhat nervously as Megan McDevitt, a former elementary-school educator turned Facebook learning development partner, called their feedback training to order.

Delivering feedback, McDevitt said, wasn't simply encouraged at Facebook, but required. If you saw something that could be improved, you were obligated to speak up, even if that meant pulling aside your boss, or their boss, for an uncomfortable conversation. "We expect feedback to go in all directions," she said. "If that conversation is with someone in a higher-level position than you, we expect you to have it. Hierarchy doesn't matter in this situation."

For the next four hours, McDevitt drilled the group on the fundamentals of delivering feedback within Facebook. If someone was holding up a project, or starting to micromanage, or not giving you a voice in meetings, that called for a feedback conversation. There was no wrong time to bring these things up either. At Facebook, you could always pull someone over and say, "Hey, I have some feedback for you." This feedback class, taken by more than 40 percent of Facebook's employees, has helped institutionalize the behavior.

Facebook's method for sharing feedback is adapted from the training company VitalSmarts. It has three main components: (1) State a fact; (2) share your story; (3) make an ask. The **fact** is the objective description of what happened. For example: When we last spoke, you said you'd have an answer to my question within a few days, and it's now been two weeks. The **story** is the explanation that's developed in your mind for why the thing you didn't like happened. For instance: I know there's a good chance you've been overloaded with work, but I've told myself you might disagree with the direction of my project, and therefore you haven't answered. The **ask** is a question meant to get to a resolution: Can you help me understand?

When McDevitt called people up to simulate these conversations, the room grew tense. Telling someone they are deficient, even in a role-play, is not easy, it turns out. (I sat toward the front of the room and wanted to dissolve into my seat during the first simulation.) But with practice, the encounters grew smoother.

After an emotionally taxing day, McDevitt wrapped by telling the group they were now required to initiate a difficult feedback conversation, preferably within twenty-one days, and asked everyone in the room to note it to themselves.

"No one said we had to commit!" one participant protested.

"I'm telling you now, it's time to commit," McDevitt replied. "These conversations don't actually have an impact if we don't hold them. That's the unfortunate science behind this. So you need to make a commitment."

The room laughed uneasily. But there was no debate. Everyone put pen to paper.

Facebook the Vulnerable

McDevitt trains Facebook employees on how to deliver feedback, but these classes make them amenable to accepting it too. When you take the training, you see feedback at Facebook isn't intended to break people down, but to expose them to new viewpoints. This can mean discussing a problem, or simply listening when someone says, "Hey, I have an idea and here's why we should try it." Ego and fear make these conversations difficult in most organizations. But within Facebook, the class, along with a broader commitment to feedback, has made them almost normal.

For Zuckerberg, this feedback culture functions similar to Bezos's six-page memos. By instilling a belief in his employees that all colleagues are worth listening to, Zuckerberg ensures ideas for new products rise up in Facebook, no matter their origin, and often come straight to him.

This idea pathway is critical for Facebook, the most vulnerable of all the tech giants. With no popular operating system of its own, Facebook has nothing to keep its users coming back other than their interest in its products. If it can't keep people intrigued, it will shrink and then die. Every few months, there's another headline reminding us of Facebook's precarious position, touching on everything from declining teen usage to daily active-user slowdowns to drops in sharing among friends. "Facebook is seriously at risk," Mark Cuban told me. "People don't need it." To stay alive, Facebook must therefore rapidly invent.

In a return visit to Menlo Park in September 2019, I sat down with Zuckerberg again and led off our conversation by asking what would happen if Facebook stopped inventing. I thought it was a straightforward question. But Zuckerberg laughed, searched unsuccessfully for a reply, and then, true to form, asked me to answer instead.

"What do you think?" he said.

"It wouldn't be good for Facebook at all," I replied.

"Yeah, obviously."

"It would just sort of fall apart," I continued.

"The idea of stopping is not even something I would think about," Zuckerberg said, still trying to find his footing. "It's a funny question."

For Zuckerberg, contemplating stasis was baffling given that he's designed Facebook's culture to churn out inventions, and to do so quickly. Facebook, he said, aims to release products as soon they're ready, even if they're not 100 percent polished, and then get feedback and revamp. This is why Zuckerberg has made feedback a priority internally, even as he often ignores feedback from the outside, an oversight that regularly throws Facebook into crisis.* The company invents, improves, and invents some more. At Amazon, it's always Day One. Facebook, as the internal saying goes, is 1 percent done.

"'Move fast' is often maligned these days because people have interpreted it as 'Just do something and don't care about the consequences.' That was never the intent," Zuckerberg said. "But this is basically the nature of it: How do we learn as quickly as possible?"

Rapid invention is Facebook's bounty and curse: Its ability to create and adapt has kept it relevant amid many challenges, from user fatigue to computing shifts. But it's also released new products faster than it can control them, often overwhelming itself. When the company hasn't "moved fast" to address its products' problems—ahead of the 2016 US presidential election, for instance—it's led to disaster. For Facebook, mitigating this system's downsides will be crucial to its long-term sustainability, as important as releasing new things.

"Iterating quickly in a direction without openness and scrutiny

---

* Facebook has a long history of ignoring feedback from the public. In its early days, the company stood steadfast as its users organized protests of features like the News Feed and a stand-alone Messenger app. After Facebook stuck to its guns and those products succeeded, it started to discount public complaints as a matter of practice. This has led to all manner of crisis for Facebook. It's been caught disregarding user privacy, playing loose with personal data, seeming unconcerned about violent content on its platform, being oblivious to foreign election manipulation, among many other flare-ups. If Zuckerberg paid a fraction of the attention to outside feedback as he does to inside feedback, Facebook would be in a much better place.

of the direction you're going in will eventually take you in a bad direction," Zuckerberg said. He knows the consequences well.

Building a Feedback Culture

Zuckerberg's obsession with soliciting feedback is natural given his background. Unlike Bezos, Pichai, and Nadella, Zuckerberg hadn't held any other job before his current one. When he started Facebook in his Harvard dorm room in 2004, he had no idea how to run a company. And when he dropped out, he learned by asking people who had.

For nearly fifteen years, Don Graham, the former owner of the *Washington Post*, has counseled Zuckerberg. The two met in 2005, connected via one of Zuckerberg's Harvard classmates whose father worked at the *Post*. When they first met, Zuckerberg didn't seem to know the difference between revenue and profit, Graham told me. He was young, green, and running a company of six people at the time. But as the company grew, Zuckerberg kept calling Graham, who was happy to advise someone he viewed as receptive to feedback. Graham proposed an investment in Facebook (which Zuckerberg turned down after another, better offer came in), and eventually joined the company's board.

"Mark is a listener," Graham told me. "I've certainly seen occasions when Mark stood against most of his advisers in doing something they didn't feel he should do. But I've also seen occasions where Mark changed his mind about something he had strong feelings about. He learns."

In 2006, Zuckerberg called Graham with an unusual request. "He called me, this was very rare, and said, 'I now realize that we've grown to the stage where I'm a CEO and I have to think of things other than what I've thought about all my life'—which was code and whatnot—'and I would like to come and shadow you for three days,'" Graham recalled. "And I thought to myself, that was ridiculous. That is the dumbest thing I ever heard of in my life. I explained to Mark that my role as a CEO could not have been more different from his. But he said, 'No, I'd like to come along.'"

Zuckerberg made the trip, and walked around with Graham, almost completely unrecognized, absorbing the inner workings of one of the world's largest newspapers. "I took him out to see our press run. You talk about a total analog experience, that was old technology. Newspapers being printed and put into trucks. So it wasn't his world. But he was watching the relationships among people," Graham said.

Two years after Zuckerberg shadowed Graham at the *Washington Post*, he called again with another request, this time to see if Graham would be willing to introduce him to Jeff Bezos, whom he wanted to shadow too. Graham forwarded the request to Bezos, but in the years following Zuckerberg's inconspicuous visit to the *Post*, his stature had grown significantly. Bezos, who would later buy the *Washington Post* from Graham, called to respond. "Well, that would be an interesting thing to do," he told Graham. "But, Don, other than being shadowed around by Angelina Jolie, I couldn't do anything that would bring my life to a complete standstill more than have everyone see Mark walking around."

I asked Graham if Bezos and Zuckerberg are similar in any way.

He said yes, both were open to new ideas, even wild ones, no matter their origin. "I brought Jeff an idea from out of deepest left-center field, the idea of buying the *Washington Post*, and I didn't do any selling," Graham said. "He got it, from never having thought about it in his life."

Pathways

Zuckerberg listens and learns, but he's also decisive. His feedback culture ensures people and ideas are not constrained by hierarchy. But Zuckerberg is not running an organization without hierarchy. When he makes a call, Facebook goes all in. This makes the channels through which ideas bubble up to Zuckerberg central to the way Facebook functions. And there are four main ways ideas get to him: his Friday Q&As, Facebook's internal groups, his inner circle, and his product reviews.

Zuckerberg's Friday Q&As date back to the days when the company occupied a single room, back in 2005, where they were simply called Friday Hangs. "We'd get Chinese food, hang out, relax," Naomi Gleit, one of Facebook's longest-tenured employees and its VP of product management, told me. The company now live-streams these Q&As, holds them in a large cafeteria, and orchestrates them with a moderator.

Zuckerberg holds the Q&As to get a pulse of the company. He wants to know "what people are thinking about, what's on their minds, what kinds of questions they're asking, what the tone is," Facebook HR head Lori Goler told me. This opens the door for anyone to bring up ideas for what the company should invent next. "They might ask about a product strategy and, in the course of asking, say, 'Here's my feedback on this product—what are you thinking in terms of strategy?'"

Facebook's employees are also constantly chattering in hundreds of internal Facebook groups, where they discuss products, ask questions of other teams, and rate their executives' performance. These groups help ideas bubble up to Zuckerberg and his lieutenants, who spark and participate in the discussions themselves. Seeing the commercial value of this internal social network, Facebook has turned it into a product called Workplace, which now counts Walmart, Domino's, and Spotify among its customers.

Zuckerberg's inner circle plays a significant role in channeling ideas to him as well, and he's tried to fill it with people who speak uncomfortable truths (though not always successfully, as we'll see shortly). Facebook's leadership team holds *Give and Take*, a book by Wharton professor Adam Grant, in high regard. The book places people into four categories: agreeable givers, disagreeable givers, agreeable takers, and disagreeable takers. The categories are straightforward. Agreeable people are liked; disagreeable people are not. Givers give to the company. Takers take from the company.

Agreeable givers don't fill out Facebook's top ranks, Gleit told me. "One thing Mark's talked about, and that we've talked about as a leadership team, is that some of the most valuable people in your organization are disagreeable givers," she said. "We really try to protect those people. I've seen Mark surround himself with disagreeable givers. They're not going to tell you just what you want to hear. They're going to tell you what they really think."

This explains why Zuckerberg has kept Peter Thiel, the controversial venture capitalist, on his board. "A lot of people would not want Peter on their board because he is such a contrarian, and Mark did," said Don Graham, who sat on the board with Thiel for years. "Peter became a director because he was a very early investor. But Mark wanted him to stay because Peter was such a loud voice putting forward ideas that Mark disagreed with."

Zuckerberg, still only thirty-five, has filled his inner circle with more experienced people, seeking to learn from them. This is where Sheryl Sandberg fits in. When Zuckerberg was twenty-three, he realized he needed someone to help him grow the company's business, and so he reached out to Sandberg. At the time, Sandberg was Google's vice president of global online sales and operations, had worked in the Clinton White House, and received CEO offers from companies in Silicon Valley. Zuckerberg had complete command of the company until that point, but offered Sandberg control of Facebook's advertising, policy, and operations divisions, hoping to get her on board. After a gut-check call to Graham, who had tried to get her to work for the *Washington Post* after Clinton left office, Sandberg joined Facebook and has been its COO since.

Sandberg has helped grow Facebook into a multibillion-dollar business, and the company would not be where it is today without her. But she's also been at the center of the company's recent scandals. Facebook users don't trust the company in part due to her ad sales team's thirst for data. And her team's willingness to take Russian rubles for US political ad buys during the 2016 election remains one of the most puzzling decisions in tech history. The name of Sandberg's conference room, Only Good News, is odd given that she helped establish the company's feedback class.

With Sandberg looking after Facebook's business, Zuckerberg has focused (perhaps a little too much) on creating new products and services, spending large parts of his afternoons in meetings with his product managers, reviewing their work, and making calls on what they should pursue. These product managers' feedback plays a significant role in determining the company's direction. "Zuck, internally at least, has a well-earned reputation of being influenceable," Mike Hoefflinger, a former Facebook director and author of *Becoming Facebook*, told me.

From a business standpoint, Facebook's feedback culture would prove vital as a major computing shift threatened to upend the young social network.

Facebook's Day One

In 2011, Facebook was in trouble. The company had built a well-functioning website, but its mobile apps were buggy, slow, and a major liability. People were starting to access the internet on mobile devices instead of desktops. And as they spent more time on phones, Facebook risked losing their interest, and hence its relevance.

Facebook's apps languished primarily because the company refused to adapt its development practices for mobile. When building its desktop website, Facebook released new features rapidly, then looked at the data, tweaked, and released again. It could update its site an unlimited number of times each day, as loading each new version took only a refresh. But when building mobile apps, Facebook was subject to the lengthy review processes of iOS and Android, leaving it much less flexibility.

As app usage rose, Zuckerberg tried to shoehorn Facebook's desktop approach into mobile by building a mobile website and adding a "wrapper" of native code for iOS and Android. The wrapper allowed the site to appear as an app, list in the app stores, and still update multiple times a day. But the hybrid product's performance suffered, and someone needed to set Zuckerberg straight.

That someone was Cory Ondrejka. After one of Zuckerberg's Friday Q&As, Ondrejka, a Facebook mobile engineering vice president, pulled him aside and said the company needed to rethink the way it operated to succeed on mobile. Instead of trying to preserve the old way, Ondrejka argued, Facebook should build natively instead, going in on code that would run on the operating system. To do this, Zuckerberg would need to accept that Facebook would have less freedom to iterate. But it would also give its apps a chance to function properly.

"I told him the current path wouldn't get us there so we needed to change course," Ondrejka told me. "Change was going to be really difficult, but I knew the new approach would work."

Zuckerberg was willing to test Ondrejka's ideas and gave him a small team to develop an experimental native app. A few months later, Ondrejka's experiment started performing better than Facebook's web-based app. And when Zuckerberg reviewed it, he couldn't deny reality, and pivoted the entire company toward building native apps.

"I'm sure my reaction was like, 'Are you sure? Can we pressure-test this a little bit more?'" Zuckerberg told me. "But over time, then it's like, 'Okay, well, if that's really true, it's just a dramatic change in the plan of the company and what we now need to go do. Let's figure out what that's going to mean.'"

Building native apps meant a significant change in the way Facebook operated. The company had to rethink the pace at which it released new features, moving from multiple times a day to every two months (that window eventually shortened, and is now almost back to normal). It had to reimagine the way it hired, looking for native-app developers after previously screening them out of its recruiting process. And it had to train its existing pool of engineers to build for native operating systems. In August 2012, Facebook released a native iOS app that was faster and less buggy than the web-based app. A similarly improved Android app debuted four months later. The rebuilt apps put Facebook on a much better footing, but Ondrejka wasn't done.

Amid this development process, Ondrejka came back to Zuckerberg with some more feedback. He drew a curve showing how quickly Facebook's users were adopting mobile, and where the company's mobile usage was heading—up and to the right. Further change was necessary.

"I looked at the growth curve and extrapolated it into the future with a mild acceleration. And it was one of those curves you look at and go, 'Well, there's no way we're going to hit that.' But if we're even close to that curve, mobile is going to be more than half of everything pretty quickly," Ondrejka said. "We were never below the prediction. The mobile transition happened even faster than the crazy graph that I drew."

Looking at this graph, Ondrejka advised Zuckerberg to dissolve Facebook's dedicated mobile team and make the entire company develop for mobile instead. Zuckerberg came around quickly, and told his product managers they'd only be able to bring him demos on mobile devices from then on. Show up with a desktop mock-up alone and they'd be kicked out of his office. It was a turning point for Facebook. The company's mobile experience improved dramatically, and today more than 90 percent of Facebook's advertising revenue comes from mobile.

The myth about Facebook's mobile transformation is that Zuckerberg had an epiphany and brilliantly repositioned his company for the age of the smartphone. This isn't quite right. The real story is Zuckerberg set up a feedback culture. And when people bought in, they brought him ideas—tough ideas that required rethinking how the company operated—and those ideas ultimately saved Facebook from disaster.

From Broadcast to Private

Facebook survived the shift to mobile, but the company encountered another dangerous moment a few years later when its most important product, the News Feed, grew stilted and dull. In Facebook's early days, the News Feed was vibrant, unruly, and unpredictable. When you opened it, you could find anything from wild party pictures to offbeat status updates from friends to people you knew trying (and failing) to flirt discreetly.

But as Facebook grew—in part thanks to how well it worked on mobile—the News Feed changed. People kept connecting with each other, transforming their networks from small collections of friends to amalgamations of nearly everyone they'd met in life. And as their networks grew, people began to self-censor. They didn't want everyone they'd ever encountered seeing their true selves.

As people built larger networks on Facebook, the News Feed algorithm also had far more posts to consider. It prioritized those generating the most engagement, and began displaying the best parts of people's lives: engagements, weddings, and babies. Amid the milestones, people grew even more reluctant to share casual posts, fearing they'd look frivolous. And so by 2015, people were sharing fewer original posts on Facebook, and the News Feed was a shell of its old self.

Facebook's executives realized this was a major problem and went into action to fix it. "We were seeing that the feed was becoming more pressurized," Fidji Simo, the head of the Facebook app, told me. "People during our research were telling us, 'Yeah, I feel less comfortable sharing than I did two years ago.' That's definitely a warning sign that you need to innovate and figure out a solution."

To stay relevant, Facebook had to wind back the clock. It had to give people a chance to share with smaller, more targeted groups of people, even though its network was now more than 1.5 billion users. Once again, the company had to change everything.

The first element of Facebook's transformation materialized naturally. People, too intimidated to share things with everyone they knew, began sharing more in Facebook's Groups—targeted networks for people with similar interests. New parents, for instance, felt more comfortable asking parenting questions to groups of new parents than their entire friend lists. So they started sharing there instead.

"There was a lot of enthusiasm starting in 2015, 2016," Simo told me regarding Groups. "The reason for that was simply adoption was picking up. It wasn't that we were doing such different things. But it was just, people were adopting the product like crazy."

With Groups membership climbing by tens of millions each month, Facebook started pushing them hard. It built new tools for Groups organizers, set high internal goals for "meaningful" Groups membership, and began selling Groups in its public messaging. Posts in Groups showed up in their members' News Feeds, restoring some of the feed's vibrancy, and making Facebook feel like a comfortable place to post again. "Groups definitely add some vitality to the app and to the feed," Simo said. "Absolutely."

Though Groups injected a crucial dose of life to the News Feed, they didn't address sharing among friends and family, which was Facebook's bread and butter. And that type of sharing was starting to gravitate elsewhere.

"The Most Chinese Company in Silicon Valley"

At around the same time Facebook was working out its News Feed issues, an upstart messaging app called Snapchat—led by the brash Stanford graduate Evan Spiegel—built a feature called Stories, which let people share photos and videos with friends that disappeared in a day. Snapchat's users loved how Stories gave them a carefree way to post (in contrast with Facebook, where your posts would go to everyone and stick around forever), and the app's usage exploded. Spiegel, who once spurned a $3 billion acquisition offer from Zuckerberg, was now hitting him where it hurt. In the zero-sum game of social media, where time spent on one platform is time not spent on another, Spiegel had the energy, the sharing, and was driving his company toward a hot IPO.

As Snapchat took off, an eighteen-year-old developer named Michael Sayman joined Facebook. Sayman had built a game that caught Zuckerberg's eye, and the company hired him as a full-time engineer in 2015. Sitting through orientation, Sayman heard speeches about how Facebook's leaders would listen to anyone's ideas, and took the message to heart. "I believed it," he told me. Before orientation was over, he spun up a presentation about how teens, already drifting to Snapchat, were using technology, and how Facebook might want to build for them.

Still barely old enough to buy a lottery ticket, Sayman started presenting his ideas to Facebook's executives and soon found himself in front of Zuckerberg. His presentation didn't initially impress. But Chris Cox, then Facebook's head of product, convinced Zuckerberg to give him a small team to experiment. "There was no blueprint," Sayman told me. "I had a few ideas, people thought that they should let me be creative, they gave me the head count to be creative, and there was no problem."

As time went on, Sayman watched his fellow teens sharing less on Facebook's family of apps and more on Snapchat. He turned his focus to Snapchat Stories, which he believed Facebook should build into its products. "I wanted the company to feel like Snapchat was an existential threat," he said. "I wanted Facebook to panic."

Sayman brought his concerns to Zuckerberg, who had heard from others who came to similar conclusions. As a teenager, Sayman was invaluable. He could help Zuckerberg learn Snapchat's culture. "He would point us to, 'Here's the media that I follow,' or 'Here are the people I think are influential, who are cool,'" Zuckerberg said. "I'd go follow those people, or talk to them, have them come in. That ends up being the iterative process of learning what matters."

Zuckerberg said he followed these tastemakers on Instagram, and confirmed he's a Snapchat user too. "I try to use all the stuff," he told me. "If you want to learn, there are so many lessons out there where people will tell you about things you're not doing as well as you could. People tell you so much if you just care about understanding what they're looking for."

This sort of experimentation has led Zuckerberg to some unexpected places. "When we were originally thinking about formally building a dating service for Facebook, I signed up for all the dating services," he told me. "I was showing [my wife] Priscilla one of the apps. It was an app where you got matched with one person a day. I was like, 'Here's this app.' And she said, 'Hey, I'm having dinner with her tomorrow night!'" He matched with his wife's friend. No word on how that dinner went.

Sayman confirmed Zuckerberg was a willing Snapchat student. "He'd send snaps with me and I would critique him on his snaps," Sayman said. "I'd be like, 'No, Mark. That's not how this works!'"

Eventually, the groundswell of support for Stories within Facebook—generated by Sayman and others—got through to Zuckerberg. And in August 2016, Facebook's executives called reporters into their offices to reveal a new product they called Instagram Stories. The product was a carbon copy of Snapchat Stories, lifting everything including the name. "They deserve all the credit," then–Instagram CEO Kevin Systrom told TechCrunch, nodding to Spiegel and his team.

Copying Stories was ruthless. It slowed Snapchat's growth considerably, and likely destroyed billions of dollars of value in its parent company, Snap Inc., which is trading below its IPO price as of this writing. Snap, frustrated and weakened, is now speaking with FTC's antitrust investigators about Facebook's anticompetitive tactics, relying on a dossier it's built up called "Project Voldemort," a reference to the Harry Potter villain.

Evil villain or not, Facebook would've been in serious trouble without Stories, which helped it recapture the friends-and-family sharing it was losing a few years ago and restore the vibrancy to its app. Facebook is still losing teen users in the US by about 3 percent each year, according to eMarketer. But without Stories and a renewed focus on messaging (another form of intimate sharing), it could've been in a much worse position. Copying was a move of self-preservation.

Sayman credited Facebook's ability to stay relevant to an internal awareness of its place in the world. "Facebook is just an internet app.

Especially in 2015 and 2016, it was just an internet app. Any other app could come about and beat it," he said. "Mark was like, 'What do people want? Let's just give it to them.' He was a bit more cautious. He was more vigilant. He was definitely not thinking his product was eternal."

In China, where copying and iterating on products has long been the norm, Facebook is known as "the most Chinese company in Silicon Valley," according to Chinese venture capitalist Kai-Fu Lee, who wrote about this in his book *AI Superpowers*. I sat down with Lee on one of his periodic visits to the Bay Area, and asked him to share his feelings about Zuckerberg. "Why do we stigmatize copying?" he said. "Don't we learn everything from copying first? Don't we learn music by copying Mozart and Beethoven? Don't we learn art by copying whichever style that is taught? Through copying, you understand the essence of what you're building, then you can innovate and build. It would seem copying is a reasonable way to get started."

From the moment Facebook copied Stories, it's iterated on it and improved it. And now its version is widely considered better than Snapchat's. Some of Facebook's improvements have been so good that Snapchat has even copied them back.

The graveyard of dead social networks is littered with the corpses of companies that were once unstoppable but were ultimately done in by pride or an inability to invent. Myspace, LiveJournal, Foursquare, Friendster, and Tumblr are among them. Facebook, meanwhile, has reinvented repeatedly and remains on top, in large part due to its feedback culture.

"Obviously, we'd rather be genius and invent first," Lee told me. "But if you can't, then copy first and then iterate."

Enter the Machines

In early 2012, Gil Hirsch and Eden Shochat, two Israeli entrepreneurs, walked into Facebook headquarters for a check-in. Their company, Face.com, was licensing its facial-recognition technology to Facebook for use in its "tag suggestions" feature, and their counterparts in Menlo Park wanted to talk. This feature, still active today, identifies who's in a photo and nudges people to tag them.

When the two entrepreneurs arrived on campus, they walked to a conference room for a meeting with Facebook's product team—or so they thought. Much to their surprise, Zuckerberg walked in instead, and he started pelting them with questions. Facebook, at the time, was unable to build features like tag suggestions on its own, because identifying faces in photos required expertise in machine learning that Facebook did not have. Hirsch and Shochat, meanwhile, were applying computer vision brilliantly within Zuckerberg's own product, and he was eager to learn more about what they were doing. "Zuck was curious from the get-go," Shochat told me. "He knew that something interesting was going on there, and he wanted to be close to such technology."

For the next ninety minutes, Zuckerberg interrogated Hirsch and Shochat about the future of computer vision and facial recognition. And as the conversation wrapped, his focus turned to acquisition. "If it makes sense, we should make this work," he said before walking out. Six months later, Facebook bought Face.com for at least $55 million.

When Facebook's engineers got their hands on Face.com's technology, they began to understand machine learning's potential, and

Facebook's executives decided to make a significant, long-term investment in the technology. At that moment, Zuckerberg began recruiting Yann LeCun, one of the world's preeminent AI researchers.

In spring 2013, Zuckerberg came to LeCun with a proposal. Join Facebook, he said, and the company will build you an AI research lab where you'd be free to pursue any AI research your heart desires, as long as you help Facebook apply it from time to time. LeCun, who lived in New York, said he'd sign on if he could remain in place and keep teaching at NYU. Zuckerberg agreed, LeCun accepted the deal, and Facebook went from AI novice to a world leader in corporate AI research almost overnight.

"There're three or four people who have done the seminal work in the last couple of decades in AI," Facebook chief technology officer Mike Schroepfer, who was in on the pitch, told me. "We were able to get Yann LeCun, who's one of them."

As all this was taking place, Joaquin Candela, a researcher who once taught a machine-learning class at Cambridge University, was tucked away in Facebook's advertising division, using his expertise to predict when people would click on ads. Candela liked his job, but when Facebook brought LeCun on board, new opportunities emerged for a person with his background. The company needed someone who could apply LeCun's research to Facebook's products, and Candela fit the bill. In fall 2015, Facebook named him director of applied machine learning, a newly formed group tasked with putting LeCun's research into action.

When I first met Candela in June 2016, less than a year into his job, he looked at me straight-faced and said, "Today, Facebook could not exist without AI." I nodded politely, and didn't believe him. But now, three years later, I do. Without AI, Facebook would not be able to manage the vast amount of execution work it needs to support its products. And the example of Facebook Live is illustrative.

In December 2015, Zuckerberg's product team rolled out Live, a feature that would let people broadcast live video on Facebook with the tap of a button. The feature made posting video on Facebook easier than ever, opening the door to a wide array of new content. Some of Live's early videos were delightful, including one of a woman laughing uncontrollably while wearing a Chewbacca mask. But some, inevitably, would be equally bad. In *BuzzFeed*'s newsroom, not long after Live debuted, one of my colleagues wondered aloud about where this new product could lead. "Someone is going to get shot on this thing," he concluded.

It didn't take long. In February 2016, just three months after Facebook Live debuted, a Florida woman named Donesha Gantt went live on Facebook after being shot in her car. "Mama I'm bleeding," she said. "I know they shot me, but it's good. It's good. God forgive me for all my sins. God forgive me for everything."

Following Gantt's shooting, Facebook Live aired videos of graphic violence at a rate of about twice per month. Murders, rapes, child abuse, torture, and suicides all appeared on the service. And these videos spread fast, tapping into humanity's morbid curiosity. The live-streamed suicides were particularly jarring, raising worries they could inspire copycats, a horrifying possibility given Facebook's size and influence on young people.

As these problems came into focus, Zuckerberg called me in for that initial meeting. The fifty-seven-hundred-word manifesto he was about to publish contained designs for a more interventionist

Facebook, one in which the company would step in more often to protect its users from hate speech, terrorism promotion, graphic violence, and bullying.

With humans reporting and reviewing this content, Facebook could only address so much. "The current system is that people report content to us," he said. "We review more than a hundred million things a month, so we're looking at a lot of content. We've built out a big team to go and look at this content. But there are billions of things that are posted on Facebook every day. If you include messages and comments, it's tens of billions of things that are posted every day. I don't think it would be physically possible, no matter how many people we employed, to go look at that content. The only way that we can really do this is by building artificial intelligence tools."

Zuckerberg's suggestion that AI could proactively flag and review posts was more than theory. Before our conversation, he had directed Candela's team to figure out how to do it. Facebook has the most extensive data set of human behavior ever to exist. It knows who we are, what we like, what we do, and how we act when something is going wrong. This data set is similar in nature to Amazon's twenty-year collection of purchase data. Just like Amazon can run that data through its machine-learning systems and figure out what we're about to buy, Facebook should be able to run its data through its systems and figure out when a video is about to air violence or self-harm.

As Zuckerberg spoke about these systems, he cautioned that the AI could not yet do this work on its own. "The theme is empowering people," he said. "When people think about AI at the limit, they're thinking about computer systems that do all these things. What's actually going to happen in the nearer term is once we get an AI system, it's not going to be perfect—there will be issues—but it will hopefully be good enough that it is worth flagging what it finds to people."

As we spoke, Facebook's AI systems were already proactively detecting some of the content Zuckerberg hoped they'd catch and passing it along for review. Working alongside these systems, Facebook's moderators turned into auditors, similar to Amazon's vendor managers. Facebook's AI looked at more content than any human could, and decided if it merited intervention. The AI also reordered moderators' queues, putting posts that *really* needed action up top. The moderators would then review these decisions and determine if the AI had made the right call.

For these systems to work effectively, Facebook needed to get the inputs right too, and Candela's team built more tools to give the company's AI team a chance to shape them. These tools, with names like Cortex and Rosetta, helped Facebook employees instruct the AI systems on what types of posts to look out for. They could note the keywords and behaviors in these systems, which could proactively look for posts with similar attributes.

With such tools at their disposal, a single Facebook employee's impact increased exponentially. Instead of sitting back and waiting to review posts flagged by humans (often flagged by people without knowledge of Facebook's policies), they could diagnose the characteristics of posts that merited action, and instruct Facebook's AI to look for them amid the tens of billions of things posted each day.

Zuckerberg grew particularly impassioned when he spoke about suicide. "It's hard to be running this company and feel like, okay, well, we didn't do anything because no one reported it to us," he said.

"When it seems like someone might hurt themselves or commit suicide, we want to help flag that for people so they have the tools to go reach out to the person or get them the help they need."

Less than a month later, Facebook announced it was widely rolling out an AI-based suicide-prevention tool. This tool, the company said, was already more accurate than humans in reporting instances in which Facebook's team needed to send help.

About a year later, Guy Rosen, Facebook's VP of product management, posted an update on how the overall program was performing. He said Facebook's AI was helping the company proactively review posts containing nudity, hate speech, and graphic content. It was also automatically removing terrorist propaganda (nearly two million pieces in a single financial quarter). And it was alerting Facebook's moderators to people considering self-harm, to whom they sent first responders more than a thousand times.

"When you're talking about people committing suicide on Live, it's incredibly hard to tell yourself, 'I'm going to be able to separate that out from all the good that's happening on the platform,'" Facebook app head Fidji Simo told me. "The fact that our AI is able to detect these kinds of things, give people hope in real time, flag it to local authorities, and save lives—that's night and day. First, there's a lot of impact just in that. And it allows us to have more confidence in having the product out there for all the good use cases."

Candela's boast was correct. Facebook today couldn't exist without AI. Without it, terrible posts would overrun the company's products, paralyzing its product teams and consuming its leadership. Facebook's AI systems are still far from perfect. And some of its moderators work in miserable conditions, as recent reporting revealed. But these systems should get better over time, both as the AI improves and as Facebook, now under pressure, improves its moderator conditions.

With the assistance of these tools, Facebook's employees can focus on creating what's next, and its leadership has the bandwidth to consider these new ideas and keep bringing them to life.

Robot Compensation

Algorithms, AI and otherwise, have reduced execution work inside Facebook so effectively that the company's human resources division is using them to determine how much to pay its employees. "Our compensation is entirely formulaic," Lori Goler, the company's HR head, told me. "The combination of your assessment and the company's performance determines your bonus, and your increase in pay, and your equity grant, and all of that."

Facebook built its compensation system in the early 2010s, when its human resources team figured algorithms could be more efficient and less biased than humans. Managers and employees tended to waste time working through compensation issues. Managers with discretionary pools of raise money might also disproportionately reward people similar to them, introducing bias. A uniform raise system based on performance ratings still wouldn't be perfect. Without well-thought-out evaluation criteria—and Facebook has overemphasized growth metrics—it could lead to monomaniacal focus on work that generates a high rating. But if implemented right, it could assure raises were handed out with as little variance as possible.

"We took all the discretion out of that system because it's the discretion that leads to bias in the organization and leads to unfair outcomes and differences in data across gender and race," Goler said. "Once you take out the discretionary aspect of it, what you're left with is much more objective."

Facebook's algorithmic compensation model put the individual performance rating at the center. These ratings, handed out on a five-point scale ranging from "doesn't meet expectations" to "redefines expectations," are plugged into the algorithmic system, which considers them alongside company overall performance to determine pay.

Performance ratings are determined every six months, when Facebook goes through a review process for both its individual employees and the company as a whole. During that process, employees get feedback from everyone they work with. Managers then read that feedback, assign a rating, and take it into "calibration sessions," where they talk through each of their reports' ratings with their colleagues and adjust as necessary. These sessions are meant to ensure each person's rating is delivered fairly.

At the end of these sessions, a final rating is determined, and the number goes into the system, which spits out the comp figures. These figures are final. "You can't ask for more money," Goler said.

This compensation technology further cuts down execution work and makes room for ideas. "You don't want every day to be a new day for you and your team members to be talking about compensation," Goler said. "You sort of want it to happen once a year when you're promoted, and that's it. The rest of the time you're focused on the work."

New Inputs

Early in the morning of April 10, 2018, I walked into a large hearing room in the Hart Senate Office Building in Washington, D.C. The room was filled with reporters, many of whom I recognized from gatherings in San Francisco, and its public viewing gallery was packed too. The reporters sat at long wooden tables, crammed shoulder-to-shoulder, on what was clearly a busier day than usual. After looking around, I took a seat, opened my laptop, set down my coffee, and prayed it wouldn't spill.

The room buzzed with idle anticipation. Senators filed in and scrolled through their phones. People in the gallery looked around, taking in the scene. And the reporters checked Twitter. Finally, Mark Zuckerberg walked in.

It had been a rough fourteen months for Zuckerberg since our initial meeting in Menlo Park. In that time, Facebook revealed it had missed a large-scale Kremlin-sponsored misinformation campaign on its service during the 2016 election. And further reports showed that Cambridge Analytica, a data analytics firm, illicitly used millions of Facebook users' data in its work for Donald Trump's presidential campaign. These events damaged Facebook's credibility, hurt its standing in the world, and landed Zuckerberg a date with the Senate Judiciary and Commerce committees.

As Zuckerberg walked in, I wondered how this man, so driven by feedback, so determined to find out what others were thinking, could have been so blind to vulnerabilities in his service. If my colleague had been so sure that Facebook Live would air shootings, why hadn't Zuckerberg anticipated them? If it was clear that Russia had engaged in a widespread campaign to undermine the US democratic process, why had he said it was a "crazy idea" that misinformation on Facebook could've impacted the 2016 election outcome? And why had he seemed so caught off guard when the Cambridge Analytica reports hit, spending multiple days in silence before responding?

The answer is an essential lesson about the nature of feedback systems. Though Zuckerberg asks people for feedback, the simple act of asking is itself insufficient. Feedback systems, just like machine-learning systems, are only as good as their inputs. And while Zuckerberg surrounded himself with disagreeable givers—people who told hard truths that helped improve Facebook's product and grow its advertising business—they were almost all techno-optimists, believing Facebook's work was "de facto good" and spending little time thinking about what could go wrong. The disagreeable giver Andrew "Boz" Bosworth, a Facebook executive, said it best in "The Ugly," his June 2016 post on Facebook's internal group, as my colleagues Ryan Mac and Charlie Warzel and I reported for *BuzzFeed News.*

> *We talk about the good and the bad of our work often. I want to talk about the ugly.*
>
> *We connect people.*
>
> *That can be good if they make it positive. Maybe someone finds love. Maybe it even saves the life of someone on the brink of suicide.*
>
> *So we connect more people.*
>
> *That can be bad if they make it negative. Maybe it costs a life by exposing someone to bullies. Maybe someone dies in a terrorist attack coordinated on our tools.*

> *And still we connect people.*
>
> *The ugly truth is that we believe in connecting people so deeply that anything that allows us to connect more people more often is de facto good. It is perhaps the only area where the metrics do tell the true story as far as we are concerned.*

After we published this post, Bosworth said he wrote it to inspire debate. And Zuckerberg disavowed it. "Boz is a talented leader who says many provocative things," he said. "This was one that most people at Facebook including myself disagreed with strongly. We've never believed the ends justify the means."

Whether Boz was trying to be provocative or not, his post was evidence that people inside Facebook weren't thinking enough about the ways bad actors could exploit their products. Their overwhelming optimism was palpable in often awkward interactions between skeptical reporters and the company's product managers. When Facebook would bring us in to introduce a new product, its representatives would be so bullish on the change-the-world nature of their inventions, they'd speak in a tone bordering on condescension. "New stickers in Messenger are going to make the world a more communicative and expressive place," they might say, "and we are so very excited to get these stickers into the hands of our users and watch the amazing things they'll do with them." Meanwhile, the Kremlin was figuring out how to manipulate their News Feed, Groups product, and ads platform.

As Zuckerberg sat before the Senate, he came close to acknowledging this gap in Facebook's feedback system. "Facebook is an idealistic and optimistic company. For most of our existence, we focused on all the good that connecting people can bring," he said in his opening statement. "But it's clear now that we didn't do enough to prevent these tools from being used for harm as well. That goes for fake news, foreign interference in elections, and hate speech, as well as developers and data privacy. We didn't take a broad enough view of our responsibility, and that was a big mistake."

Zuckerberg, understanding that his feedback system needed new inputs, began adding them. To fix the system, the company started hiring ex–intelligence officials, journalists, academics, and adversarial-minded media buyers, and told them to pressure-test its systems.

"You're looking for a mentality," Justin Osofsky, Facebook's VP of program management and global operations, told me. "The person is passionate about finding, identifying, understanding, and addressing risks before they happen."

A few days before the 2018 US midterms—the first big test of Facebook's ability to stand up to further election manipulation—I met with James Mitchell, Rosa Birch, and Carl Lavin, three people at Facebook who've seen the integration of these new "inputs" on the ground level. Mitchell heads Facebook's risk and response team, which works to find vulnerabilities in its content-moderation systems. Birch is a program manager in its strategic response team, which coordinates Facebook's response to crises across divisions. And Lavin is a former editor at the *New York Times*, *Forbes*, and CNN, who works on the company's investigative operations team, a group formed entirely to think about the bad things people could do with Facebook's products.

Mitchell and Birch work hand in hand with the more adversarial-minded people Facebook has brought in, Lavin included. (Fun story: I once tried to work for Lavin. He did a deep Google search on my background, found a correction on an article I'd written for my college newspaper, and never emailed me again.) "We need people internally thinking about these things, not to only react to things that come to us from advocacy groups or journalists and public officials," Lavin said.

As someone who's covered Facebook for years, it's a bit weird to imagine ex–intel people and ex-reporters working alongside the company's product managers, but it's clear they're bringing different thinking to the organization. "It's great to be able to have conversations about threats and risk and people say, 'Well, here's how we think about threats: we talk about the capabilities and the motives of the actors, and we talk about the vulnerabilities,'" Lavin said. I had never before heard the words *threats*, *vulnerabilities*, and *motives* uttered in Menlo Park.

Speaking of Menlo Park, Facebook has made it a point to hire people outside it, seeking to get away from the homogeneous thought and techno-optimism prevalent in Northern California. "We don't actually have lunch together because most of us are not in California," Lavin said. "We're in Dublin, Singapore; I'm in Austin, Texas. That's purposely to give us a non-California focus on the world."

To inject these adversarial thinkers' views into its veins, Facebook has paired them with longtime employees who understand the ins and outs of its product and process. "If you don't know how it's going to manifest on the platform, then you can't necessarily translate an issue externally to what we might see internally," Mitchell said. "We try to bring both [types of people] into the fold because both become particularly important when we're trying to understand how systems might be used and abused."

The groups come together in a variety of forums, including incident review meetings, where the company's product, policy, operations, and communications teams get together each Friday and dive into the company's mistakes. Sitting in the room gives Facebook's new, less optimistic employees a chance to speak and bring up things that otherwise might've gone unnoticed. "It's the people, but it's also the new process," Mitchell said. "Without the process of that type of a review meeting, we're just coming up with stuff, and it doesn't matter what we're coming up with."

On top of these formal settings, Facebook's new team members are infusing their feedback into the company's internal groups. "It's hugely important," Birch said of the chatter in these groups. "It enables rapid communication to be really easy, takes enormous noise out of your inbox, and it really helps bond teams together, especially when they're not in the same location."

Mitchell's, Birch's, and Lavin's teams also play a significant role in helping Facebook's machine-learning systems know what to look for. To address hate speech spreading in Myanmar (where critics have accused Facebook of helping to spark a genocide), Birch's team enlisted Facebook's machine-learning engineers to build a system that could pick up the language. Then they fed it keywords, image attributes, and other red flags to help it determine what to send to moderators. These tools made the new moderators Facebook hired to deal with the crisis more effective, but in this case, much of the damage was already done. Facebook is now working to head off potential crises in other countries, including Cameroon and Sri Lanka, and it now has the infrastructure to act faster.

Gaps in Zuckerberg's feedback system have led to years of chaos for Facebook. But that very system may help the company recover quickly. Facebook's employees are now actively listening to the "new inputs"—the ex–intel officers, journalists, media buyers, and other adversarial thinkers—who are finding a receptive audience in a company filled with people trained to consider others' ideas. "There's an eagerness to receive this information," Lavin said.

When our meeting wrapped, Birch walked over to Lavin and made some small talk, asking when he'd head back to Austin. The plan, Lavin replied, was to stay in Menlo Park until the following Tuesday—election day. The 2018 midterm election came and went. Lavin stuck around. Though Facebook will inevitably face more tests in the future, it passed this one. No significant exploitation of its service emerged.

Facebook's Next Reinvention

On September 25, 2019, Zuckerberg took the stage at the San Jose McEnery Convention Center, standing before the words "The next computing platform." He was there for Oculus Connect, a developer conference for the operating system and hardware bundle Facebook is building for virtual reality.

Oculus represents Facebook's effort to become something more than an app. By building its own operating system for what it believes will be the "next computing platform," Facebook is hoping to free itself from the source of its vulnerability: being subject to its competitors' whims.

"There are things that you can only do when you build the platform," Zuckerberg told me. "We build apps on phones, and we build websites. And especially in the model of apps on phones, we are often quite restricted by what the operating system makers think that apps should be able to do."

Zuckerberg expressed frustration with the way mobile operating systems make you pick a task first and then the person you want to do it with. You tap a messaging app and *then* tap the person you want to message, he said. This is the opposite of human nature, where you first pick a person and then the task.

"One of the things that I hope to do—and that I hope we do in AR and VR—is influence the direction of the next computing platform to be more focused on the organizational principle, making it around people, rather than just tasks," he said. "This is an area where I really care about the direction that computing goes in."

Zuckerberg knows what it's like to depend on others, and he doesn't want to do it forever. If virtual or augmented reality takes off, Facebook, through Oculus, will have its own popular operating system, giving it a say in the way people interact with its services, the likes of which it doesn't have on desktop, mobile, or voice. Zuckerberg wants this badly, and is investing in Oculus to set the stage for Facebook's next big reinvention.

Facebook does not lack for ambition or the technology or processes necessary to keep its uncommon success going well into the future. And as I walked out of Zuckerberg's glass-walled conference room, this seemed more evident than ever. If the company can manage its growth in a healthy way—by listening to the new inputs and behaving responsibly—it will be a force for decades to come. If it can't, now that federal regulators are bearing down and politicians are calling for its breakup, Facebook will end up in the very place Zuckerberg has long sought to avoid: as a footnote in technological history.

CHAPTER 3

INSIDE SUNDAR PICHAI'S CULTURE OF COLLABORATION

In July 2017, a little-known Google engineer named James Damore wrote a ten-page memo critiquing the company's diversity and inclusion practices. He composed his memo after attending Google's anti-bias trainings, and sent it to the sessions' organizers in an attempt to deliver some feedback.

Unequal representation of men and women in tech, Damore wrote, may be in part due to biological differences, and not overwhelmingly due to bias, as the trainings emphasized. Women are more neurotic than men, he said, a potential reason why they hold a smaller percentage of "high-stress" jobs.

"At Google, we're regularly told that implicit (unconscious) and explicit biases are holding women back in tech and leadership," he said. "Of course, men and women experience bias, tech, and the workplace differently and we should be cognizant of this, but it's far from the whole story."

When the trainings' organizers didn't respond, Damore shared his memo with "Skeptics," a small internal Google email group filled with people who don't readily accept the status quo. The group, one of thousands of email communities constantly buzzing at Google, seemed like a natural place to discuss the document. But after Damore sent the memo, its members began circulating it with others inside the company, and it spread rapidly.

Damore's memo soon became the talk of Google's internal communications networks, and it divided the company. Some Googlers discussed the merits of his arguments. But to a greater extent, they debated whether Google should fire Damore, and whether his supporters should go too. Hundreds of his colleagues wrote to him following the memo, mostly in approval, he told me.

As the debate raged, someone shared the memo with Gizmodo's Kate Conger, who, while on vacation with spotty cell service, published it. The story reached millions of readers, transfixing a public growing steadily more uneasy with the treatment of women in the workplace (the Me Too movement would ignite just two months later). In an instant, what began as a post on a small, internal Google network became a top global news story.

Amid the chaos, Google CEO Sundar Pichai, on a trip abroad, was forced to make a decision. He could keep Damore on and risk having Google's employees feel he was condoning a suggestion that women were neutrotic and this neuroticism was keeping them from leadership roles. Or he could fire Damore and risk sending a message that the free expression so valued within Google wasn't that free after all.

In a note to employees, Pichai made clear that while he welcomed dissent, Damore's implication that women were less biologically suited for Google's work crossed a line. "Our coworkers shouldn't have to worry that each time they open their mouths to speak in a meeting, they have to prove that they are not like the memo states, being 'agreeable' rather than 'assertive,' showing a 'lower stress tolerance,' or being 'neurotic,'" Pichai wrote.

He fired Damore.

The Hive Mind

Ideas move fast inside Google—so fast that those advancing them regularly lose control. And that's by design. Sometimes, this means conversations traditionally confined to the watercooler become international incidents. But the communication tools through which Damore's memo spread have also turned Google into one of the most collaborative companies on earth, linking its employees in a collective consciousness and breaking down the typical boundaries between divisions. With the help of these tools, and Pichai's guidance, Google has reimagined itself multiple times, weathering a series of computing shifts that threatened to sideline it.

Google's dominance may seem inevitable—the story of a company that cracked the search code and rode it to an $800 billion market cap. But in today's fast-transforming business world, Google hasn't kept pace by milking a single product. It's repeatedly reinvented itself—and search, in specific—to keep up with changing consumer preferences, and its success can be traced to its ability to do so.

Google search has gone through many evolutions: It began as a website, but after Microsoft cut its distribution on Internet Explorer, it reinvented itself as a browser, Chrome. Then, when browsing moved from desktop to mobile, Google reinvented again, placing search at the heart of a mobile operating system, Android. Now, as people operate mobile devices with their voices, Google is reinventing search in a voice assistant.

With each reinvention, Google builds elements of its existing product suite into a new offering, requiring intense collaboration. The Google Assistant, for instance, places Google search, Maps, News, Photos, Android, YouTube, and more into one cohesive product. To build these products, Google needs to be able to work seamlessly across groups. And its array of internal communications tools—both custom-built and publicly available—make this collaboration possible.

Google's employees work entirely inside Google Drive, for instance, using Docs, Spreadsheets, and Slides to write plans, take meeting minutes, store financial information, and deliver presentations. Across the company, files in Drive are almost universally open, so Googlers working across groups can read up on ongoing projects and figure out how they've evolved, where they're headed, how they make money, and who's doing what. This makes Google unprecedentedly transparent for a company its size.

"At a company that large, having that level of access and transparency made it super easy to do a lot of research on your own and help connect yourself with the right people," one ex-Googler told me. "Everything was available internally. You could search through all the corporate documents."

When a Googler identifies someone they want to work with, they can study up and connect with them via Google's intranet, Moma. "There's an entire company directory where you can visualize everyone's reporting structure, and within that you can see their headshot, their email address, access their calendars, and book time on their calendars," the former Googler told me. "That was the biggest piece, being able to easily find the right people when we were trying to do things outside the realm of our daily work."

By operating with an open Drive, Google also sparks collaboration inside the documents themselves. While he was working on his first presentation within Slides (Google's version of PowerPoint), Matt McGowan, a former Google head of strategy, watched with surprise as his colleagues jumped in and began adding to it all at once. McGowan initially stepped back from his laptop, afraid to add anything else. But he later discovered his team did this purposefully, as a way of introducing him to Google's culture, and he quickly embraced it. "I'm sitting there at home one night and I have team members of mine all around the world all digging in, adding information," McGowan told me. "Things built really quick because of that."

Because Googlers work inside Drive, there's an unwritten rule against attaching documents to email, saving them from working in multiple versions of the same document at once, and inevitably reconciling them. "That eliminates version control," McGowan said. "Think about the time you save when you eliminate those issues." When you factor in Drive's search capabilities—which intelligently suggest documents based on when they were created, how often they're accessed, your relationship to the person who created them, and other signals—these tools help Googlers inform themselves about their colleagues' work and contribute to it, with great speed.

Googlers also connect with one another via email distribution lists like "Skeptics," using them to talk about everything from work projects to things with little application to Google's business. "You could join any of these distribution lists—I don't recall a moderator," Jose Cong, a former Google head of talent, told me. "Most of the topics are what you would expect: people discussing ideas, people asking for help on technology challenges, support groups. There was a group of cyclists that was sharing tips on how to take rides around campus. By the time I left, there was a document where people were openly sharing their salaries."

These lists help information and ideas travel rapidly inside Google. "Because the tools are there, the digitalization is there, the connectivity is there, it's a lot easier now to share opinions that have been shared for decades," Cong said. "Before they happened at the coffee machine, they happened over lunch. In today's age, there is this ability to not have to do it at the coffee shop—to do it broader."

The company also comes together monthly for question-and-answer sessions with its leadership called "TGIF." These sessions are held at the Google's Mountain View campus, inside a large cafeteria called Charlie's, and can feature an update from Pichai, a presentation from another executive or team, and then some questions.

TGIF is technology enabled too. Googlers from around the world can tune in for a broadcast of these sessions via the company's intranet and ask questions via Dory, a Q&A software tool named after Dory from *Finding Nemo* (a fish that suffers from memory loss and asks lots of questions). Inside Dory, Googlers vote on questions they want answered during the Q&As, and can do so without seeing the other votes, so the crowd won't influence them. Management typically answers the top ten questions. When I visited Google's campus in February 2019, I saw vote counts on Dory ranging into the thousands.

Finally, Google has its own internal social media tool called Memegen, a website where Googlers post memes reacting to storylines within the company. When I visited Google, I saw memes praising Pichai for his testimony before Congress, joking about the company's promotion criteria, engaging in bathroom humor, mourning the loss of a colleague, and apologizing for an email that accidentally went out to the entire company. (When Marissa Mayer left Google to become CEO of the struggling Yahoo!, the top Memegen post was a picture of her accompanied by the text: "Accomplished tech leader, Finally leading a non-profit.")

"That's where I went to see employee sentiment," Cong told me. "By seeing what they're creating, you get an idea of how things are trending."

Google's communication tools are critical to its success: They cut down the execution work required to get up to speed on a new project, and make room for new ideas. They send ideas rocketing around the company, sparking invention and improvement. They enable collaboration and signal it's expected, removing red tape and driving home the importance of working together with fellow members of the hive mind.

These tools have helped Google reinvent search repeatedly over the past fifteen years. And through each evolution, Sundar Pichai has been critical.

The Toolbar Episode

Pichai joined Google as a product manager in 2004, amid a developing crisis. At the time, approximately 65 percent of Google's search traffic originated from Microsoft's Internet Explorer browser, leaving the company fairly exposed. The fiercely competitive Microsoft wasn't likely to send billions of dollars of search traffic to another company forever, and Google's leadership (reasonably) feared Microsoft would try to replace Google with a search product of its own.

To fortify itself against Microsoft's power, Google built a number of products—including Google Toolbar and Google Desktop—that would give people access to its search outside of Internet Explorer's default settings. Google Toolbar, for instance, put a big Google search field underneath Internet Explorer's address bar, making Google a prominent part of the browser for those who installed it.

Pichai, an understated, lanky engineer who grew up in south India without a phone or a refrigerator, took charge of Toolbar with a clear directive: get it on people's computers. The experience set in motion his rise to the top of the company.

When Pichai took over Google Toolbar, the product had gained traction among some early adopters who appreciated the easy access to search. (Until then, the best way to search on Internet Explorer was to click a "Search" button that opened up a search web page.) It also blocked pop-up windows, winning it more supporters. But years into the project's life, Toolbar didn't have nearly enough downloads to fortify Google against Microsoft. So Pichai began developing partnerships to force the issue.

"The hardest part about getting someone to try a new piece of Windows software is getting them to download it," Linus Upson, a Google VP who shared an office with Pichai at the time, told me. "So he built relationships with Adobe, which had the most-downloaded Windows products on the planet in Flash and Acrobat Reader. When you got Flash, or you got Reader, there was a check-box that said, 'Would you like Google Toolbar?' And he did this with a number of popular downloads at the time. He set up a distribution channel."

In the meetings with Adobe and others, Pichai needed to figure out how to bring people with disparate interests together, often in tense negotiations, where lots of money changed hands. It helped that Google had lots of money at its disposal, a product of its cash-printing ad business. Tempting as it might've been to show up, flash the money, and tell Google's partners what to do, Pichai listened instead, validating their opinions, and worked toward solutions.

Pichai's demeanor during the Toolbar episode previewed the way he'd encourage invention as he worked his way up Google's ranks and into the CEO role. Bezos channels ideas up to decision makers via six-pagers. Zuckerberg creates a direct path for ideas to get to him via his feedback culture, making sure they flow up and down. Pichai makes sure ideas flow side to side, breaking down barriers between groups, setting objectives but minimizing his presence, and sparking collaboration.

"Sundar's not the kind of person that dominates the conversation. He does a very good job of making space for other people to have their thoughts heard," Upson said. "He's very thoughtful, very deliberate; he's very good at listening to other people."

As Google Toolbar grew, Microsoft made life difficult for Google. "There would be a weekly fire we were fighting. Numbers would dip, and we would suddenly notice something was wrong, and we'd have to figure out what happened," Aseem Sood, a former Google senior product manager who worked directly under Pichai, told me. "Eventually, we had to get the Justice Department involved to make sure Microsoft knew we were paying attention."

Two years into Pichai's tenure at Google, Microsoft released IE7, a foreboding Internet Explorer update. Within months, Microsoft switched the default search on Internet Explorer from Google to its own Live Search, a predecessor to Bing, bumping Google from a spot that was once its lifeblood.

Pichai's distribution deals saved the company from a potentially catastrophic scenario. In his first test at Google, he grew Google Toolbar to hundreds of millions of users, generating billions of dollars in revenue and fortifying Google against a vicious attack. But Google's war with Microsoft was only heating up.

The Path to Chrome

Pichai interviewed at Google on April 1, 2004, the day the company released Gmail. Google had a history of outlandish April Fool's pranks, and Pichai wasn't quite sure whether this new email service, whose gigabyte of free storage far exceeded that of other web-based email services, was real or a prank. He spent his first set of interviews trying to figure it out.

"People kept asking me what do you think of Gmail? I hadn't had a chance to use it. I thought it was an April Fool's joke," Pichai said in 2017. "It was only in the fourth interview when someone asked me, 'Have you seen Gmail?' I said no, and so he actually showed it to me. And then the fifth interviewer asked 'What do you think of Gmail?' and I was able to start answering."

Gmail was no joke. It was the first prong in Google's assault against a core element of Microsoft's business. Microsoft, at the time, was making good money selling Office, a set of productivity programs that included Outlook for email and scheduling, Word for word processing, and Excel for calculations. To use these tools, you'd pay Microsoft and install them on your computer.

Starting with Gmail, Google introduced its own version of every one of Microsoft's core productivity programs—all of them on the browser, which Google leadership correctly believed would be the future. In March 2006, Google acquired Upstartle, a company whose Writely product became Google Docs. In April 2006, it introduced Google Calendar. And in June 2006, it introduced Google Spreadsheets. Combined with Gmail, these tools posed a credible, browser-based challenge to Office, sending Microsoft into check.

Google's attack left Microsoft with an interesting choice: It could continue to improve Internet Explorer, its market-leading web browser, which would make Google's web-based tools faster while imperiling Microsoft Office. Or it could inch IE forward ever so slightly, hoping to keep its top spot while holding Google (and the web) back. Microsoft chose the latter.

"Microsoft was motivated to have Internet Explorer be good

enough so that it kept its number-one position, but not good enough so that it would make web apps like Gmail better experiences than things like Outlook," Chee Chew, a former Microsoft GM who moved to Google in 2007, told me. "Microsoft at the time largely defunded Internet Explorer staffing. It took the team down to essentially maintenance mode."

With Microsoft impeding Internet Explorer, the browser became slow and bloated. This wasn't received well by Google leadership, who saw Microsoft attacking its search business and hampering its productivity tools as well. That Microsoft carried out this attack with a poorly functioning browser made it susceptible to challenge.

Google at first significantly invested in Mozilla Firefox, Internet Explorer's top competitor. But at a certain point, the company decided its ideal browser needed to be built from the ground up, and that Google ought to build it itself.

"From a purely technical standpoint, we came to the conclusion that we wanted to start from a blank slate, just throw all the legacy away," Upson told me. "There are times when it's best to start from scratch."

Google thus embarked on a project to build a new browser with a clear goal: speed up the internet. If this browser caught on, it would give Google's web-based apps a much better chance to succeed. It would also reinvent the way Google provided search. Instead of having to download Google Toolbar, or navigate to Google.com, people could type their queries right in the browser's address bar, freeing Google from the whims of its competitor.

Google named the browser Chrome, a tongue-in-cheek reference to its goal to minimize the browser's "chrome," a term for anything that's not where the browsing occurs, such as the address bar, tabs, buttons, and widgets. To lead this initiative, Google leadership again called on Pichai.

Building off his successful run with Toolbar, Pichai took an unconventional approach at the helm of Chrome. To foster collaboration, he created a decentralized organization that operated with the spirit of an open-source project. Pichai's team invented Chrome in a fashion similar to McGowan's Google Slides presentation, a collaborative endeavor with loose central decision making. Pichai gave the groups working on Chrome a directive to make the browser fast, simple, and secure, and then handed them ample leeway to build things into the product.

"Sundar wasn't the gatekeeper. It wasn't that you had to go to Sundar," Chew, who worked on Chrome versions 1 through 44, told me. "For the vast majority of things that we did in Chrome, we didn't talk with Sundar. We didn't go and get it approved by Sundar. He and Linus [Upson] created a culture where people were empowered deep in the organization."

Members of the team would still consult with Pichai, who would advise them on what might be best for the project. But when I pressed Chew on who would make the final decisions to push things through, he told me there was no such person. "You have to step away from the canonical way of how companies operate," he told me. "Suspending disbelief, suspending everything you know, probably helps you more than trying to put it in the framework of what you do know."

In his work on Chrome, Pichai took the freewheeling culture created by Google founders Larry Page and Sergey Brin to its natural end. He pushed his power down to the rank and file, giving them a chance to make their own decisions on how they should build. Instead of inserting himself as a bottleneck, he got out of the way and let his team work. The people working with Pichai rewarded him with quite a few good ideas. Chrome ran each tab as if it were a separate program, so if one tab broke down, the entire browser wouldn't crash, as was typical among Chrome's competitors. Chrome also brought search and web navigation together into a single address bar, simplifying an experience traditionally kept in two separate fields. And the browser ran fast, as intended.

Chrome's open-source development style worked so well that Google decided to open-source its code to the public upon launch, via a developer version of the browser called Chromium. "Our intent here is to help drive the whole web platform forward," Pichai said as he introduced Chrome in 2008. "As the web gets better, there's a direct strategic benefit for Google. We live on the web. We build services on the web. If the web gets better, more people use the web, and Google benefits."

When Chrome debuted, Pichai had to sell it to two entities: the public and his colleagues, many of whom had poured significant effort into developing Firefox. Google had invested in Mozilla, the nonprofit that built Firefox, in a deal that made Google the browser's default search engine. "The idea of going and competing with Firefox is not something any of us wanted to do," Upson told me.

Pichai won over his colleagues not by brute-forcing his product, but by letting them figure it out for themselves. He never mandated that Googlers use Chrome. "It was, 'Could we win over our employees just on the merits of the product itself?' " Upson said. "Even today, you don't see everyone using Chrome at Google. You still see people using Firefox."

This soft-handed approach won Pichai trust across divisions inside Google, and from Google's founders. "Sundar was very good at managing that room," Sood, who sat in on Pichai's presentations to Page and Brin, told me. "Not in a manipulative way. He was very genuine, he's very empathetic, he doesn't really come with ego, he's not coming in with his own ideas. He's very good at commanding that crowd."

Chrome's debut was promising but not overwhelming. The browser won over early adopters but struggled against the forces of inertia that kept Internet Explorer strong. To push Chrome's adoption into meaningful territory, Pichai tapped the distribution channels he had built with Toolbar, got Google leadership to commit to a massive ad budget, and then Chrome took off.

"The myth is, we built this browser, it was great, everyone used it," Upson said. "But the reality is, we got the first tens of millions of users who were enthusiasts, but crossing that curve into the hundreds of millions of users—those distribution channels that he built with Toolbar and we then later used for Chrome are what got us up into the hundreds of millions of users, and then the organic flywheel took off."

Chrome debuted in 2008. By 2009, it had 38 million active users. By 2010, it had more than 100 million. And today, more than 1 billion people use Chrome. Microsoft, meanwhile, has ceased developing Internet Explorer.

The Shift

Almost as soon as Chrome put Google on stable footing, the ground underneath the company began to shift again, this time with the potential to make Microsoft's Internet Explorer threat look quaint by comparison.

As Google built Chrome, advances in connectivity and processing technology made it possible to shrink computers down to the size of your hand, giving birth to the smartphone era. The iPhone and an array of Android devices (the operating system owned by Google) replaced flip phones in millions of people's pockets. And the web browser, designed for use with a mouse and keyboard, didn't translate well on mobile devices, whose small screens were poorly suited for free-flowing web surfing, and so people started tapping through the internet via apps.

The more time people spent inside apps, the less crucial Google search became. Instead of searching Google for restaurants in your area, you'd use the Yelp app. Instead of searching for flights and hotels, you'd use Kayak. Instead of searching for news and information, you'd have it pushed to you by Facebook and Twitter. Google search was built to sift through the open web's limitless number of pages and show you exactly what you wanted with the guidance of a few keywords. On mobile, its raison d'être wasn't immediately apparent.

Another significant technological advance arrived around the same time. After years of floundering, artificial intelligence researchers began making breakthroughs largely thanks to the same processing and connectivity advances that led to the rise of smartphones, which also generated the vast quantities of data they needed to prove out their models.

"We as an industry had finally gotten enough computational power to actually make these things work on real problems," Jeff Dean, a Google senior fellow and the head of the company's AI-research group, Google Brain, told me.

Google (along with the broader tech industry) started to invest heavily in AI after seeing promising early results. And the company, never shy to bet on research with no clear business outcome, dedicated significant resources to researching three major AI disciplines: computer vision, voice recognition, and natural language understanding. "From those three different important domains—language, vision, and speech—it was clear that there was something real there," Dean told me. "We started to see more and more results where if we scaled up the size of the model and used more training data, we would actually get better and better results."

Computers still aren't nearly as intelligent as humans, but these advances helped them interact with the world as humans do. With AI, computers transformed from glowing two-dimensional screens to things that could see, hear, process natural language, and speak back. And in November 2014, Jeff Bezos's invention factory put it all together, releasing the Amazon Echo and its embedded digital assistant, Alexa.

In Mountain View, Google's leadership took note.

The Split

On August 10, 2015, Larry Page published a shocking blog post. Google, one of the world's most recognizable brands, would henceforth be known as Alphabet. And Alphabet would be made up of a collection of companies, including Calico (a Google antiaging project), Life Sciences (a Google health research group now called Verily), and a newly refined, reborn Google.

Since its founding, Google routinely invested in projects that didn't advance its mission to "organize the world's information and make it universally accessible and useful." The mission is straightforward for a search company. But the boundless curiosity of Google's cofounders and its iconoclastic employees pushed it in many directions over the years, to the point that it became an unwieldy mix of science projects and capitalism.

With Alphabet, the founders would return Google's concentration to its original mission and break off the science projects into their own companies under the broader Alphabet umbrella. In this new organization, Page and Brin would assume the CEO and president roles, and Pichai, who was running all the company's products outside of YouTube by then, would take control of Google.*

"This new structure will allow us to keep tremendous focus on the extraordinary opportunities we have inside of Google. A

---

* Page and Brin would step down from their roles at Alphabet in 2019, handing the entire company over to Pichai.

key part of this is Sundar Pichai," Page wrote in the post. "It is clear to us and our board that it is time for Sundar to be CEO of Google."

The Alphabet restructuring confused many outside Google, but the motivations were clear inside: At the time, the mobile web was becoming increasingly less relevant, making traditional search less useful. By 2017, apps would account for 89.2 percent of all time spent on the mobile internet in the US, leaving only 10.8 percent to the open web on any browser, according to eMarketer. Meanwhile, the Amazon Echo, a device many laughed at initially, began fielding questions from people who grew to see Alexa as a friend. Answering questions was Google's turf, and Amazon was encroaching on it.

Standing still at this juncture was not an option for Google. It needed to reinvent again to stay relevant as voice computing and mobile apps transformed the way people interacted with the internet. The Alphabet reorganization set the stage for a big bet, one that could be made without distraction.

AI Answers

As soon as Pichai took over Google, he directed his employees to adopt an "AI first" mentality, encouraging them to build AI into their products at every opportunity. "He wanted to galvanize the entire engineering and product community at Google to say, 'Hey, there's something real happening here; we should pay attention,'" Dean told me. "That shifted the thinking of some of the teams that weren't already thinking in this direction."

Google's collaborative culture helped Pichai's directive take hold quickly. When Dean's team would build AI technology for one team, word would spread quickly, leading to increased demand for the technology from other teams, and new applications would soon follow.

When the Google Translate team, for instance, used an AI model to predict what a sentence in one language might look like as you wrote it in another, other teams took note. The Gmail team eventually used the same model to create Smart Replies, a feature that suggests short, AI-generated sentences to reply with when you get an email in Gmail.

As AI adoption took off inside Google, the company's products grew smarter, more aware of one another, and better at understanding the people interacting with them. Google Photos began to identify gestures, like hugs, and let people search for photos with them. Gmail started sharing flight confirmation details with Google Calendar, which marked them down automatically. And Google's voice search became proficient at answering questions spoken in natural language, as opposed to typed-in keywords.

As these developments picked up, Googlers' imaginations kicked into gear. A group in Zurich began working on ways to make search more conversational, other groups began thinking about how to make Google more personal and helpful, and the hardware division's attention drifted toward speakers.

The company was making progress toward something, but it still didn't quite fit together.

The Great Reinvention

In early February 2016, Sundar Pichai walked onstage at Charlie's café in Mountain View with a smile. A playful Sergey Brin had just told the company Pichai would deliver Google's 2016 strategy as an interpretive dance, kicking off the new CEO's first annual address with a line that gave the packed cafeteria a good chuckle.

The promised dance performance would've thrilled the crowd, but Pichai took a more comfortable approach. Leaning beside a lectern, he delivered his remarks in his typical fashion, with the measured cadence of a professor.

He started with a quick summary of where Google stood. More than 50 percent of its searches were now coming through mobile, he said, many of them via voice. Search was still relevant on mobile, but not in the same way Google imagined it for desktop. Pichai then put up a slide displaying the logos of Google's products to one side with a squiggly bracket beside them pointing to the word *Assistant*.

The company, Pichai said, would spend the year working together to build a digital assistant that linked Google's major products together. This assistant would turn people's experience with Google into a running, personalized conversation. They could, with their voices, ask it how long it would take them to get to work (via Maps), when their Amazon package would arrive (Gmail), and when their next meeting would take place (Calendar). They could ask for funny videos (YouTube), pictures from their vacations (Google Photos), and updates on the latest headlines (Google News). And they could ask it to search the web.

The Assistant would be Google's response to the changing nature of the internet. It would give Google an answer to Alexa—which would inhabit 72 percent of all smart speakers by year's end, according to eMarketer—and Apple's less impressive Siri. Google Assistant would have a chance to overtake these competitors thanks to Google services, like Maps and Gmail, that were already enmeshed in people's daily lives. The Assistant would get people accustomed to speaking with Google, making it natural to ask it things they once typed into a search bar. And the Assistant would connect people with apps, developed by Google and others, giving Google a chance to continue to organize the world's information and make it useful whether it was on the open web or not.

The Assistant was a grand reinvention that required a grand collaboration. For the product to meet its potential, an array of Google's product teams—Maps, Gmail, Calendar, Photos, Search, YouTube, and more—would have to neatly knit their services together, working cross-functionally in an unprecedented way, even for them. Google's AI technology would underpin it all, making it possible for people to speak to these products, and for them to speak back.

After presenting the Assistant concept, Pichai made it clear he expected the entire company to contribute. "If you're looking for this year's top priority, I'd say this is it," he said.

When the Assistant project kicked off, it didn't exactly start smoothly. "The Assistant has followed the path of many of our cross-company collaborations in the sense that, at first, it's pretty chaotic and a little bit painful," Jen Fitzpatrick, Google's senior vice president of Maps, told me. "Not everyone was always on the same page in terms of who was doing what, or exactly what the priorities were in any given moment."

To push through the early chaos, Pichai removed obstacles preventing ideas from flowing between groups and divisions. He brought the Assistant teams together, sometimes in meetings of twenty-five people or more, helping them agree on what they were building, who would build what, and what should be prioritized.

Once everyone was on the same page, Pichai applied the approach he took with Chrome to Google more broadly: With a clear framework in place, he stepped back and let the company invent together.

"We had to go beyond a single idea from Sundar and really start to pull in the collective wisdom of a broader set of people to say, 'How are we going to take this broad idea and make it into something much more concrete, much more specific?'" Fitzpatrick said. "That's really when you get into the broader, cross-company collaboration."

This style of invention was a departure from the last major Google-wide project, Google+, which took a more centralized approach and failed. "Instead of an Assistant team coming back and saying, 'Hey, this is what we want, we want all of you to do A, B, and C,' it took this bottom-up innovation in all the teams, and gave it a label," Shiva Rajaraman, a former Google product management executive, told me. "That's the secret of collaboration at Google. When these teams come together and can amplify what they do because the company's attention is on it, it works really well."

From there, things moved quickly. Google's communications tools sped up Assistant's development, helping the various teams on the project spot new opportunities, find the right people to work with, and stay up-to-date. "We don't hide stuff from each other," Nick Fox, the VP in charge of Assistant, told me. "It wasn't a secret what was happening with the assistant. It was well known and well understood, and so then people could understand how to fit in better."

As the teams worked to fit their services together, they also tackled some novel questions about the assistant itself. They discussed whether it should have a face, whether it should be called the Google Assistant or something else (Lucky was one option), and whether it should be funny. They also discussed how it should respond to sensitive topics, understanding a kid could be listening in as an adult asked questions, necessitating more discretion than otherwise would've been exercised on a screen.

There was one more wrinkle to the process: a smart speaker. For people to hold a running conversation with Google, they'd need to be able to speak with it wherever they were, including when they were away from their phones. So the team added a speaker with Assistant built in. It would be called Google Home.

Home was a potential landmine for the Assistant project. Hardware is often produced with a certain level of top-down planning. When you're on the hook for ordering a certain number of parts by a specific date to produce a certain number of products by holiday season, there isn't much room for listening to lots of ideas on how to do it. Because of this, hardware operations are typically hierarchical.

The Home wouldn't be like other hardware products, however. It would simply be a delivery device for the Assistant. Its speaker quality was important, but the real draw was the voice inside that could canvass the world's information and tell you exactly what you wanted to know, when you wanted to know it. The Home team therefore couldn't let a rigid planning mentality spill over into the rest of its operation. It would have to come to the table and collaborate, even if it wasn't standard practice in the discipline.

"Without a doubt, there's a linearity to hardware that is required to get a product shipped, but that just sets your timing," Rick Osterloh, Google's senior vice president of hardware, told me. "Certainly for Google, the right model is not a top-down-driven product-development cycle."

Osterloh, a Motorola veteran, said the approach was different from anywhere else he'd worked, and a traditional hardware veteran coming into Google wouldn't recognize it. "They'd probably have a very difficult time with it," he said. "Almost every other place that makes hardware, an example would be Motorola, is very hierarchical, very top-down driven. Your business model completely depends on predictability. It is very important that people respond when direction is set."

Life at Google is different for Osterloh, who told me people from other groups routinely email him with ideas with little regard for the chain of command. "I greatly appreciate someone taking the time to think something through about a product that might make it better," he said. "Here, there's a big idea soup, and you're trying to create the best products you can from a number of interesting technologies and concepts that people come up with."

After working on Assistant through the winter, Google hosted its annual Google IO ("Inputs/Outputs") conference that spring in Mountain View's Shoreline Amphitheatre, a local concert venue. Early in the morning on May 18, 2016, thousands of developers, journalists, and members of the public poured into the amphitheater, settling into its seating area and lawn. I took my seat among them. It was a beautiful, clear day that smelled like coffee, freshly mown grass, and sunscreen.

After a kickoff video, Pichai came onstage and immediately made the day's big announcement. "We are evolving search to be much more assistive," he said, introducing the Google Assistant. He then showed two demos, one in which someone orders movie tickets with Assistant, another in which they order food. Tellingly, both actions would otherwise be handled within mobile apps, cutting Google out of the picture entirely.

Pichai then followed with the day's most impressive announcement: the Google Home. The device, a palmable speaker with no screen, wasn't much to look at. But when Google rolled a promo video, I sat up straight in my chair. The video showed the Home playing music, updating the status of a flight, changing a dinner reservation, sending a text message, translating Spanish to English, sharing a package delivery status, answering questions about space, reading events off a calendar, finding routes to the airport, and turning off the lights upon being told "good-bye."

The video was somewhat aspirational, but Home was a real product with a clear trajectory: these actions, still mainly carried out on screens, are on their way to being carried out by speaking into the air. With artificial intelligence technology steadily improving, this was Google's first step into a world where a voice in the sky will accompany us as we work, drive, and go about our daily lives. Speaking to it will be as natural as talking with a fellow human. It is the next iteration of search, and potentially much more.

The Assistant project was uniquely Google. It was a collaboration between a large number of product groups, buoyed by artificial intelligence, and developed with the assistance of the company's communication tools. The result turned into the product that might keep Google relevant for some time to come.

Uprising

In late fall 2017, a group of Googlers began discussing a rare secretive project underway inside the company. Google, they learned, was licensing its AI technology to the Pentagon, which was using it to decipher drone footage.

The potential for the Pentagon to someday target drone strikes with Google's technology unsettled the group, and they raised their concerns with leadership. As these talks progressed, Liz Fong-Jones, a Google site-reliability engineer, learned of the project and posted about it internally via Google+. Google's workforce, not used to being kept in the dark, then dug up the project's documentation, along with some of its code, revealing its scope. The project, which the Pentagon called Maven, was worth millions of dollars, with the potential to lead to much more if the military liked its results.

News of Maven's existence didn't go over well inside Google. The company's mostly liberal employee base was already upset about its recent sponsorship of the Conservative Political Action Conference (CPAC). Now, they learned it was working on something that could eventually be used to kill people—and doing so discreetly, making matters worse.

"It came at this moment of, like, 'Oh, man, here's another thing that Google is doing that people are unhappy with,'" Tyler Breisacher, a now-former Google employee who opposed Maven, told me. "The secrecy of it pissed people off a lot."

Growing more frustrated by winter, the Googlers wrote a protest letter directly to Pichai. "Dear Sundar," it read. "We believe that Google should not be in the business of war. Therefore we ask that Project Maven be cancelled."

The letter spread quickly via Google's internal communications tools, and a thousand people signed it within a day. Jeff Dean, who's signed an international letter against the use of AI for autonomous warfare, didn't seem surprised. "A lot of machine-learning researchers have strong views on what kinds of things they want to see their research work used for. Many of them don't want to see autonomous weapons developed," he said. "They think that's a rather dangerous direction to go in for the world."

Diane Greene, then the head of Google Cloud, the division that brokered the deal, addressed the letter at a subsequent TGIF in remarks that came off as unprepared. And when the question-and-answer session arrived, Google's employees went off. "Hey, I left the Defense Department so I wouldn't have to work on this kind of stuff," said one Googler, according to a coworker who recounted the meeting to *Jacobin* magazine. "What kind of voice do we have besides this Q&A to explain why this project is not okay?"

Things spiraled from there. Anti-Maven memes filled Memegen, thousands more signed the protest letter, and about a dozen Googlers resigned. Someone then leaked the letter to the *New York Times*, and an ensuing leak put damning quotes from emails sent by Fei-Fei Li, Google Cloud's chief scientist for AI, on the *Times*'s front page. "Avoid at ALL COSTS any mention or implication of AI," Li wrote in one email discussing Maven's positioning. "Weaponized AI is probably one of the most sensitized topics of AI—if not THE most. This is red meat to the media to find all ways to damage Google."

Watching the mounting unrest—propelled forward by Google's collaborative culture and its communications tools—Pichai listened once again. Google, at the time, was working on a framework to govern the way it developed AI. When the Googlers expressed their concerns over Maven, Pichai brought them into the process. He wanted their opinions not only on how Google should approach AI for warfare, but on other tricky ethical situations as well.

"It was his suggestion that we get a lot of input," Kent Walker, Google's senior vice president for global affairs and chief legal officer, told me. "He wanted to make sure there was widespread input from people across Google in the way we thought about this."

Google then held a series of town halls across its global offices. The town halls covered issues regarding transparency (How transparent should medical AI technology be?), when humans should be in the loop versus when it was okay for machines to operate autonomously, and whether it was okay to develop technology that would harm people.

"We were basically saying these are pretty complicated areas. It's a new and fast-evolving technology. We want to be careful, thoughtful about how we engage and what kind of factors we look at before we do anything," Walker said. "The fact that people are focused on a specific issue is a spur to make sure you get it right, to make sure you're hearing from everybody. But fundamentally we're here for the long term. We want to make sure that we're setting the right foundation for the work that everybody at the company is doing for years to come."

Hundreds of employees attended each town hall, and a series of lively discussions took place. "I actually appreciated those," said Breisacher, who eventually quit due to reservations about the company's direction. "I went to one in my last week there. I was like, 'Well, I already put in my resignation; this isn't going to change my mind.' But I was impressed that it was a really good discussion talking about the questions that were raised in a nuanced way."

Walker and his team took the input from the town halls, put a rough set of standards together, and ran them by Google's AI leadership and those who raised concerns. Then they went to Pichai, who gave feedback, and they refined it until he was comfortable enough to share it.

On June 7, 2018, Pichai released the AI Principles, the framework the company had been working toward. The AI Principles contained worthy but uncontroversial aims like "Be socially beneficial," "Avoid creating or reinforcing unfair bias," and "Be accountable to people." But Google also listed the applications of AI it wouldn't pursue, which included "Technologies that cause or are likely to cause overall harm" and "Weapons or other technologies whose principal purpose or implementation is to cause or directly facilitate injury to people."

The language felt somewhat squishy to Google's fiercest critics. And it's unclear how the company will interpret its own words in time (What, for instance, does "overall harm" mean?). But the uprising over Maven, one that used the power of Google's own collaboration tools to register discontent with management, resonated. Google said it would not renew its contract with the Pentagon.

The Maven episode wasn't the first instance of dissent within Google. The company's employees regularly register their objections. But it did indicate that Google's communications tools could be used much more powerfully in service of protest than they had in the past. And sure enough, something much bigger was right around the corner.

Exodus

A few months after the Maven protests, about twenty thousand Googlers walked off the job in offices across the globe. The Walkout, as it's now known, took place a week after an October 2018 *New York Times* article reported that the company had paid Android founder Andy Rubin $90 million upon his exit from Google following sexual misconduct accusations. The article further reported that Google had protected others accused of similar behavior.

If Maven could've been written off as an aberration, a rare moment when Googler dissent tapped into broader political movements and spilled out in public, the Walkout made clear the company was entering new territory. The communications tools that enabled Google's workforce to invent breakthrough products were now showing another side. Wherever these tools have existed for long enough, decentralized dissent movements and polarization have followed. Now Google was experiencing these forces for itself.

The Walkout began much like the "networked" protest movements of Tahrir Square, Occupy Wall Street, and the Women's March: with a flurry of social media activity from previously unknown actors.

While the target this time was a company, not a dictator or a broken political system, the origin story shares many of the same characteristics.

When the *New York Times* story broke, many at Google were enraged. Rubin, while engaged in an extramarital relationship with a coworker, had coerced her into performing oral sex, according to an accuser. Rubin, who denied the accusation, left the company with "a hero's farewell," the *Times* said, with a massive payout and a public note from Larry Page wishing him well. "I want to wish Andy all the best with what's next," Page wrote. "With Android he created something truly remarkable—with a billion-plus happy users."

Upon reading the story, Claire Stapleton, a YouTube product-marketing manager, was floored. To her, the surprise wasn't just the fact that it had happened, but that it had happened inside Google. "Juxtaposed with a culture that aspires to be something better, I was just shocked," she told me.

Throughout the day, Stapleton followed the reaction on an internal Google Moms list, where her colleagues anonymously shared their own stories of harassment, breakdowns in the HR reporting process, and discrimination. And after reading these messages throughout the day, she decided to take action.

In an email to the moms group she's since recounted to *New York* magazine, Stapleton floated the idea of mass action. "I wonder how we can use our collective leverage. . . . If we banded together, what could we do?" she wrote. "A walkout, a strike, an open letter to Sundar? Google women (and allies) are REALLY rage-fueled right now, and I wonder how we can harness that to force some real change."

The next day, Stapleton created a new Women's Walk group meant to coordinate the collective action, and shared it with the moms group. "I immediately knew we were onto something because as people poured into joining this group, they started loudly and proudly expressing their outrage on this list," she said.

When members of the group started proposing demands to take to leadership, one of its members created a Google Doc to keep track of them. In true Google fashion, the document filled up with dozens of people adding their own demands and commenting on others. In addition to the Doc and email lists, the burgeoning movement used an internal Google site to keep colleagues up-to-date, and Google spreadsheets to keep track of contact information. The Walkout organizers (now there were many) did this all out in the open, on Google's tools, with their real names.

"We really owned that this was a Google moment," Stapleton said. "There is something to be said about how the culture creates the space for us to do this. Because we're so used to debating things on Google lists, and on the internal version of Google, people are constantly registering their dissent. That's a good aspect of the culture."

With the 2018 midterm elections around the corner and momentum at their backs, the group running the Walkout decided there was little time to wait. And so they called the Walkout for that Thursday, less than a week after Stapleton created her group. "It was a beautiful collaboration," she told me. "It reminded me of how type A and good-student-achievery people are, particularly when they come together around a common purpose, at Google."

As with Maven, Pichai listened. In a note to employees ahead of the Walkout, he apologized for the company's flat response at the previous TGIF (the Google leadership team delivered a perfunctory "sorry" and moved on to a presentation from Google Photos), and said it was important to him that the company take a harder line on inappropriate behavior. He then told the employees they'd have the support they'd need during the Walkout, and that he was taking their ideas to heart. "Some of you have raised very constructive ideas for how we can improve our policies and our processes going forward," Pichai wrote. "I am taking in all your feedback so we can turn these ideas into action."

By the day of the walkout, Stapleton's group contained about two thousand members, and she and her fellow organizers had no idea how many would participate. They called the walkout for eleven a.m. local time at all offices, which set the stage for a "rolling thunder" of action. The first walkouts took place in Asia, where offices in Japan, Singapore, and elsewhere participated with significant numbers. And then Europe, New York, and Mountain View followed. By the end of the day, twenty thousand people turned up, ten times the number of people in Stapleton's group. At each location, Googlers shared stories of mistreatment via bullhorns or their voices—it all happened too fast to apply for official permits.

"People were just so flush with the excitement and the collective power even though nothing had been accomplished yet," Stapleton said. "It was this huge undertaking and this huge manifestation of something big."

The Walkout didn't leave either side all that satisfied. Google leadership was shamed by its own employees. And it now must reckon with an unsettled workforce energized by the broader anti-Trump political movement—a common line connecting the backlashes to CPAC, Maven, the Walkout, and Damore—now challenging its employer.

For those who'd walked out, only one of their demands was met: the end of forced arbitration for current employees. Like Occupy Wall Street and the Women's March, their protest's decentralized nature left them with varied requests—including one asking for employee representation on the board—and little power to make them a reality outside the threat of further protest.

I asked Stapleton how the Walkout could avoid a fate similar to that of Occupy Wall Street, a movement that fizzled out after achieving few concessions, and she told me that wasn't the point. "We're not really purporting to be some sort of cohesive, like—the Walkout carries on in lockstep together. It isn't really that," she told me. "We have two thousand people on a LISTSERV; we have twenty thousand people that walked out, supposedly. We don't even know who those people are and how to reach them. There's actually not some sense this is an ongoing movement, really. I suspect it will be reactivated and crop up as new things happen."

As with most protest movements, especially decentralized ones powered by social media, the Google protest movement then turned messy. In our talk in Mountain View, I asked Kent Walker whether Google found these movements productive. "Overall, a culture of openness and feedback and employee engagement is tied up with the innovation we create," he told me. "We value that; we need to find a way to make it work at scale."

What Walker hinted at was that Google was reimagining how its internal communications networks should function, and the company soon released a policy discouraging political talk. Google also retaliated against Stapleton and fellow Walkout organizer Meredith Whittaker, according to the two women, and both left the company. After the episode, by the end of 2018, confidence in Pichai and his leadership team dropped by double digits among Google employees.

Google's employee movements, and Google's response to them, have caused current and ex-employees to wonder whether the company's culture can persist at a scale of one hundred thousand employees. After the Walkout, Google's leadership has struggled with this question too. Pichai and his deputies have rolled back the company's openness—slimming down TGIF and firing some activists—in moves meant to preserve the culture's good elements while mitigating controversy and protest. But it's hard to have it both ways.

Ultimately, Pichai will have to decide whether he wants to stay transparent and deal with the associated employee dissent, or close the culture and live with the consequences. For Google, more openness and debate will only help it make more thoughtful decisions in the future. It will also preserve the hive mind, which, no matter how difficult to manage at times, is the reason Google can come together and tackle projects as complex as the Assistant. Without its communications tools and associated openness, Google's name might not be a verb, but among the likes of Lycos, AltaVista, Ask Jeeves, and Excite—companies that made noise in search but ultimately couldn't adapt.

CHAPTER 4

TIM COOK AND THE APPLE QUESTION

Marques Brownlee is the type of person Apple has taken an interest in of late. Brownlee is a YouTube star, with more than ten million subscribers who regularly consume his crisply shot reviews of the latest technology products. A modern-day tastemaker, Brownlee is among a new class of influencers shaping perceptions of technology companies today. And Apple's marketing machine, a well-oiled operation now in its forty-fourth year, is well aware. Brownlee is a regular invitee at Apple's release events, and the company has given him access to its top executives, including Craig Federighi, its revered senior vice president of engineering. In return, Brownlee has largely rewarded Apple with positive reviews, an exchange the company has grown accustomed to during its reign at the top of the tech world.

All this makes Brownlee's review of Apple's HomePod surprising. In February 2018, Brownlee got his hands on the new smart speaker—Apple's long-awaited answer to Google Home and Amazon Echo—and proceeded to lambast it for nine minutes and forty seconds.

Brownlee's review began innocuously enough, with a run through the HomePod's hardware. He praised its pleasing fabric, its buttons (volume up, volume down), its power cord, its squishy feel, and its world-class sound. "I've listened to a lot of different music on these different smart speakers lately and this is, in fact, the best-sounding one," Brownlee said.

But then his tone shifted. For a product like HomePod, he said, what matters most is what it can do—and there was a lot it couldn't. After faithfully naming a few basic things the HomePod was capable of—playing music from Apple Music, reading your last text message aloud, telling you the weather—Brownlee started listing the many places it fell short. It couldn't tell the difference between two people's voices; it couldn't sync with another HomePod; it couldn't make Spotify the default music player. And from there, he just let it fly.

"You can't order products online," Brownlee said. "You can't order food online, you can't call an Uber or a Lyft with it. You can't have it read calendar events or set any calendar events. You can't set multiple timers at the same time, only one at a time, which seems like something you would do in a kitchen with a smart speaker. You can't have it make phone calls via your voice—again, that's something you have to set up on your phone and then airplay to the HomePod. You can't look up a recipe, you can't use Find My Phone. The list goes on and on. There's so many things that HomePod, when you compare it to other smart speakers, just doesn't do. So in conclusion, HomePod is a weird product."

Brownlee, with a mix of confusion and disappointment, then delivered his verdict: "In most cases, honestly, you're going to be better off buying a much smarter speaker that can do way more that sounds almost as good," he said. "Getting a HomePod right now and using it a lot just magnifies everything that's wrong with Siri."

By now, you probably have an idea why Brownlee was so baffled by the HomePod. The disappointing device was a direct product of Apple's culture, one still set up for the old way of working, in which ideas come from the top.

Inside Tim Cook's Apple, the Engineer's Mindset is nowhere to be found (even though Cook is an engineer). Democratic invention is rarely encouraged, people and ideas are constrained by hierarchy, and collaboration is held back by secrecy. Apple's internal technology, meanwhile, is years behind that of its competitors. The result is predictable: Apple is great at polishing straightforward ideas handed down from the top (a speaker), but struggles to build new, inventive products that rely on ideas from across the company (the assistant inside).

Hence the most important question we can ask about the company today: Can Apple can keep pace in our fast-moving business world without a wholesale culture change? With iPhone sales slowing and a new era of computing emerging, Apple will need to let go of its rigidity or risk the fate of the HomePod: something that looks nice on the outside but isn't quite right beneath the surface.

A Culture of Refinement

When Robin Diane Goldstein took a meeting in a San Francisco hotel a few years ago, she arrived a bit early and went to pour herself a cup of coffee. Goldstein, a twenty-two-year Apple veteran, picked up a mug from a table off to the side of the room and quickly grew frustrated. "As I put my fingers through the handle of the coffee cup, I could feel the mold line on the inside of the handle," she told me. "And my first thought was, why didn't the designer and manufacturer of this spend just another thirty seconds and finish the inside of the handle to make it smooth?"

Her second thought? "Fuck you, Steve."

"Like, you ruined me," she said, mentioning it with a smile. "Most people would never think about this, they wouldn't notice it; maybe they'd realize that the inside of the handle felt a little scratchy or there was a line there, but it wouldn't be a thing. But having spent time at Apple, I realized, 'No, this matters.' The part that you can't see, that you may touch for just a few moments, is part of the entire experience."

Goldstein's story is a revealing peek into the way Apple operates. When Jobs was alive, he would come up with ideas, and the rest of the company would refine them, making sure no mold lines, or their equivalents, remained anywhere on the product. The culture prioritized execution; it was (and still is) set up to polish ideas handed down from the top.

"There was one visionary and one dictator," one former Apple employee told me, referring in both instances to Jobs. "The dictator was in charge. He had a lot of ideas. He was very dynamic and full of energy. He was leading the troops on visions he had for the company and for the products. He thought he knew more than anybody else what the products should be and how people should use them. Because of all that, because of the charisma he brought with him, people followed."

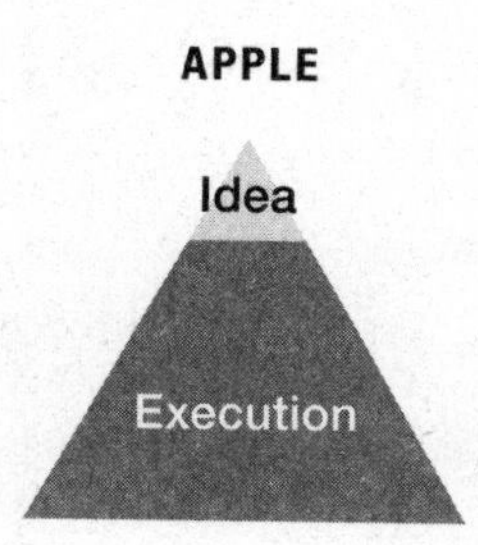

Today, Apple is still refining the two marquee products that Jobs invented before he died: the iPhone and the Mac. Apple has improved these products significantly: It's made them thinner and faster. It's made them more useful with wearables like the Apple Watch (a watch for iPhone owners) and AirPods (earbuds for iPhone owners). And it's made daily life better for iPhone owners with features like Face ID and Apple Pay (both delightful). Few companies get more out of their existing assets than Apple.

Inventing beyond these devices, however, is another story. Apple's bets to create ambitious new products—like the HomePod and its own autonomous car—are failing. And Apple's refinement culture, a relic of the Jobs era, is to blame.

The Refiner's Mindset

In place of Jobs, six Apple executives drive the company today, delivering ideas that the rest of the company executes. They are: Tim Cook, the unassuming CEO with an operations background. Eddy Cue, the colorful senior vice president of software and services. Phil Schiller, the deceptively powerful head of product marketing. Jeff Williams, the chief operating officer who oversees design. Craig Federighi, the capable and smooth senior vice president of engineering. And John Giannandrea, the Scottish ex-Googler who runs machine-learning and AI strategy. Angela Ahrendts, the former Burberry CEO and head of Apple retail, would've been among this group had she not stepped down in 2019. As would Jony Ive, Apple's brilliant and sometimes detached former head of design, who stepped down that year as well.

Apple's designers are the first line of employees tasked with carrying out these executives' orders. While engineers are royalty inside Amazon, Facebook, and Google, designers are deities inside Apple. In most companies, designers are handed something and asked to make it look nice. At Apple, designers dictate how a product will look and feel, and then it's up to the engineers and product managers to help bring it to life, no matter how technically difficult it might be to pull off.

Design's enmeshment in Apple's product development process has helped the company regularly improve its flagship devices. Apple designers do not typically hand off projects. They stay close from beginning to end, minimizing the buck passing that's typical inside companies that create "good enough" products.

"The asset that Jony brought was that he built this incredibly talented team of people that knew more than just good design," Doug Satzger, an Apple designer for more than a decade, told me. "They understood good design, good engineering, good manufacturing, business operations. Each one of those things is an experience that the user sees in the end product."

Satzger and his colleagues were so deeply ingrained in the product creation process they would regularly spend time on the production lines in China, making sure their products were meeting expectations. Some of Satzger's Bay Area–based colleagues spent as many as 240 days in China per year, he told me; others permanently relocated there.

The tradition continues today. In 2019, a leaked United Airlines document showed Apple spends $35 million on flights between San Francisco and Shanghai alone each year. Facebook, Roche, and Google, the next-largest accounts, spend "over $34 million" on all flights with United per year, a sum that doesn't eclipse a single Apple route to China.

Designers occupy such a revered place inside Apple that their colleagues are prepped on how to speak to them, sometimes down to the angles at which they show their products. "We planned it to that level of detail," one ex–Apple employee told me. "It's like every single piece from the structure of the meeting, to what information we showed, to what information we withheld, to how we phrased things, to backup plans and having other things underneath the table just in case—that's the level of detail that we spent our time on. It seemed like a lot of wasted expended effort that's not necessarily pointed at innovation. It's as if they're God. A lot of people treated them like God."

Concentrating power as such, Apple's executives have distanced themselves from the company's rank and file. Their employees are there to execute, not to deliver ideas, so there's little intentional mingling. While everyday employees at Amazon, Facebook, and Google often have stories about interactions with their CEOs, connections with Tim Cook are rare.

"I bumped into Cook and it wasn't exactly a warm interaction," Jean Rouge, a former Apple employee, told me. "I crossed him in a hallway and said good morning. He looked at me, deciding if he was going to give me any attention. He crossed me and said 'See you.' Not good morning or have a great day buddy, or anything; it was basically 'See you'—almost like fuck off, sort of like 'I don't have time for this.'

"Folks have talked about Apple culture being a little bit on the cold side—it's probably PR'd," Rouge told me. "I would call it the cold meat locker."

While Zuckerberg and Pichai hold town halls with their companies, and Bezos has his six-pagers, Apple has few channels to bring ideas to leadership. I asked the ex-employee who presented to design how she could get ideas to Cook and his circle of leaders. "Umm, probably not likely," she told me. "Probably not going to happen. I have never heard of anyone trying to do that."

Though Cook is detached from his rank and file, he's well-liked by the company's senior executives, who describe him as a thoughtful, exacting chief executive with a good sense of humor and a humble disposition.

"You can be assured that he's never going to act impulsively. He's always thoughtful about everything, whether it's some small thing or a very critical company issue or challenge," Denise Young Smith, Apple's former vice president of worldwide talent and human resources, told me. "He personally models this discipline, this excellence, this attention to detail, this constantly striving to be better and to offer the very best product you can to the customer."

In the business environment of a few years ago, where loads of execution work hampered companies' ability to develop employee ideas, you could see why Cook would be the natural choice to succeed Jobs. But today's business world is different. And Apple would likely have had to adapt regardless. A visionary who harnesses employee ideas is going to be more effective than one who doesn't.

Silos and Secrecy

Apple's product development itself takes place in extreme secrecy, and even the company's own employees are largely kept in the dark. The secrecy is meant to encourage focus, supporting the company's pursuit of excellence, and also to limit leaks.

When an Apple employee wants to speak with colleagues about a project they're working on, they must be "disclosed," or given the official go-ahead to discuss it, and their colleagues must be disclosed too. Outside of mutual disclosure situations, Apple employees are prohibited from speaking with anyone about their projects, including their coworkers, friends, and spouses.

"I couldn't talk to people on my team who needed to be doing work about the work they needed to be doing," Marc Minor, a former Apple marketing employee, told me. "I couldn't use the names of products, I couldn't use code names. You can't use the code name with someone unless they know the code name."

The disclosure system helps minimize distractions, freeing people up to obsess about their product's tiniest details. "It becomes much more explicit and clear what you as an individual should be focused on because you don't even know the other stuff," one former Apple engineer told me. "At Google, there can be a bit of a sense that everyone is responsible for everything and everyone knows everything else that's going on, and everyone's dog-fooding [internally testing] everything, and everyone's giving feedback about everything they're dog-fooding, and it results in there being less individual ownership. Whereas at Apple you only know about this one thing—and that's the one thing you have to do."

Or, as Goldstein described it: "They're siloed so that experts can be experts."

In addition to increasing focus, Apple's secrecy helps it surprise customers when releasing new products. The element of surprise hooks media and Apple enthusiasts' attention twice per year: once for new iPhone model introductions, and again for the company's Worldwide Developers Conference, also known as WWDC, which focuses on how developers can build on top of Apple's operating systems.

About a week before these events, Apple's marketing and communication teams go into a "black site," a dedicated building with blacked-out windows set aside to review and translate the new products' marketing materials. Inside, they get the materials ready for wherever they may appear, be it physical stores, billboards, or online. And then it's showtime. "It all happens in a little building in Cupertino where everyone gets locked up and they do amazing work," Minor told me, adding that he ultimately finds the secrecy worthwhile. "There's something to be said for their ability to control the message," he said. "As a marketing person, controlling the message is everything."

If Apple employees leak news of new products, or even show previews of those already announced, they are fired. When Brooke Amelia Peterson went to visit her father, hardware engineer Ken Bauer, at Apple's campus, she posted a video on YouTube that included footage of Bauer's announced but not yet released iPhone X. It was a costly mistake.

Peterson is always carrying her camera, Bauer told me. So when she filmed the phone it didn't strike him as out of the ordinary. "Seeing her with a camera, yeah, something should have gone off and said maybe this is a bad idea," he said. "But it's like if your kid loves baseball, and he wears a baseball hat every day, you just get used to it; you don't think about it."

When Peterson posted her video, it went viral, and people inside Apple took notice. "Suddenly, at eight o'clock in the morning, I get a call from Apple security—'Hey, there's a problem,'" Bauer told me. "I got a text from my boss about the same time."

Peterson deleted the video, but copies kept spreading. Bauer tried to get those copies deleted but to no avail; the internet was doing its thing. "I was devastated," he said. "The gravity of it was on me immediately. I knew right away that might be the end of my job."

In a conversation with Apple security, Bauer owned up to the mistake and tried to convince them he'd be the least likely person to make a similar one again. "It was only about a day later that it came down—'Okay, you're gone,'" he said. "They walked me out of the office and that was it, I was done."

Bauer found new employment shortly afterward—in part due to his even-keeled reaction to the publicity from his daughter's video—and he seemed at peace with the situation. "No grudge against them," he said.

These elements—the design-led development process, the focus, and the element of surprise—have all combined to make Apple's flagship devices the most desired products in the world. But these same factors are conspiring against Apple as it faces a shift of a similar magnitude to the one Google encountered when typing and clicking turned into talking and tapping.

"A Form That's Right"

On January 2, 2019, Tim Cook posted a rare letter to Apple's website. "To Apple investors," he wrote. "Today we are revising our guidance for Apple's fiscal 2019 first quarter." The letter marked the first time Apple had revised its financial predictions since 2002, when it anticipated it would miss revenue expectations by at least $150 million. This time, the miss would be much larger, at least $5 billion.

Cook listed some reasons for the downward revision, but only one mattered: the iPhone wasn't selling well. "Lower than anticipated iPhone revenue, primarily in Greater China, accounts for all of our revenue shortfall to our guidance and for much more than our entire year-over-year revenue decline," Cook wrote.

A struggling Chinese economy and the brewing trade war between China and the US played a part in the lower iPhone sales, but another factor loomed larger: Smartphones, after years of big advances, had become good enough that owning a top-of-the-line model was no longer important. People could wait longer to upgrade, and that put a dent in Apple sales. In November 2018, Apple had said it would no longer report unit sales of the iPhone, an indication of what was to come.

Apple cofounder Steve Wozniak himself offered a convincing argument that the iPhone was reaching a point where upgrading wasn't all that necessary. "I'm happy with my iPhone 8, which is the same as the iPhone 7, which is the same as the iPhone 6," he said in a 2017 interview, adding that he would not be upgrading to the iPhone X. "Look at cars. For hundreds of years a car kind of had four wheels, about the size that would fit people inside, and headlights. And so cars haven't changed that much. They reach a form that's right. And the smartphone has reached a form that fits the hand, all sizes of hands."

Cook, in an interview with CNBC, put on a brave face. When CNBC host Jim Cramer said his daughter would not upgrade her phone because she saw no need for a new model, Cook said he was fine with it. "The most important thing for me is that she's happy," he said.

But the episode brought home a sobering reality for Apple. For more than a decade, the company had focused on refining Jobs's ideas to a state of near perfection, and the party was coming to an end. The iPhone, first introduced by Jobs in 2007, had gotten slimmer and faster on its way to becoming the great consumer device of the early twenty-first century. But as Wozniak indicated, the marginal benefit of refinement was growing increasingly less apparent; the iPhone 6 was practically indistinguishable from the iPhone 7 and 8. And meanwhile, Apple's competitors were catching up, building cameras and processors that approximated the iPhone's, reducing its advantage over the field. Apple felt the loss particularly hard in China, where WeChat—an app that does chat, payments, investing, ride hailing, and more—had become the de facto operating system, allowing easy switching off from Apple's iOS.

With the refinement of Jobs's iPhone no longer a reliable path to growth, Apple has big plans beyond it, and it will need to become inventive again to realize them. But its culture—created for a visionary-led company in an era that's passed it by—doesn't appear ready to take it where it needs to go.

The HomePod Debacle

Long before the HomePod, Apple had Siri, a voice assistant it released on iPhones on October 4, 2011, one day before Jobs died.

With Siri, Apple had an opportunity to take the lead in voice computing and never look back. But to give Siri the best chance to succeed, the company needed to lose its refiner's mindset. It would have to dispense with the silos and secrecy, allowing the Siri team to mingle with other divisions to see how their products could fit in as Google did on its Assistant project. It would also have to view Siri as a product unto itself, independent of the iPhone, so it could plug in other services. None of this happened.

"Since October 2011, when Steve died, that was when the issues started," a founding member of the Siri team told me. "Cook is a very nice guy, very good at a lot of things, specifically at execution because of his operations background. But he has zero vision in terms of product."

Instead of unleashing the Siri team, Apple stuck with its silos and secrecy. It planned the team's work top-down, kept its projects secret, and made sure its members had minimal interaction with their colleagues.

The lack of collaboration—the opposite of Google's approach with Assistant—held the Siri project back, according to the founding member. "We had three badges to get into our office. And nobody else could come into our office. We were pretty much hidden from the rest of the world. Nobody knew us," he said. "They believed one team could do everything by itself. It's stupid. Collaboration is the only thing that you should have, especially when you're doing a product that uses a lot of information coming from different pieces."

Apple's decision to keep its silos and secrecy stemmed, in part, from the way its leadership viewed the assistant. Siri, to them, was an iPhone refinement: a fun, embedded personality that would make the iPhone more appealing. But this was a major tactical mistake. By focusing on Siri's personality over the assistant's utility, they produced something incompetent, causing people to lose interest.

Had Apple leadership listened carefully to their employees, they would've seen Siri as something more. The rank and file working on Siri wanted to open it up to third parties outside the Apple ecosystem, making it a voice layer on top of the web and apps, and hence very useful. But these employees got nowhere.

"For a long time, a lot of us were pushing for Apple to open up Siri to third-party developers. But they didn't want to go there," one person who worked on Siri told me. "They saw it as an iPhone feature. They didn't really see it as an operating system for the future."

Viewing Siri as such, Apple gave design outsize influence on the project, another mistake. Design imagined Siri as a magical, human-like being, multiple Siri engineers told me, which contributed to its poor performance. When the engineers sought to build a feedback tool, design rebuffed them, as asking people to evaluate Siri's performance would lessen its supernatural appearance. Without feedback, the engineers struggled to improve the assistant.

"Even if one percent of users tell you that this was wrong, or this was right, and here's what was wrong, that's immensely valuable," the former Siri engineer told me. "They didn't want to go there because it breaks the illusion or the perception that it's this person that you're talking to. I remember having a lot of fights with them about stuff like this. We can't improve stuff unless we can do things like this."

The designers also built animations into Siri that looked nice but slowed it down. When the engineers complained, the designers didn't take it well. "It was pretty hard getting that past design, because they were like, 'Oh, but look at all these beautiful animations,'" the ex–Siri engineer said.

Apple's planning process added yet another obstacle. Apple typically plans once a year, a hardware-inspired development schedule. With Siri, at least in the early going, the company would hand down a set of features to work on for the year, leaving the team little flexibility to make changes on the fly.

Apple has built great software, including its operating systems, but these systems didn't rely on machine learning at the level of a virtual assistant, which listens and talks back. An annual or semiannual planning schedule isn't conducive to building an assistant with this type of experimental technology. Such a project takes flexibility and the ability to adjust on the fly.

Operating systems, by their nature, are also containers for different software programs, which act independently inside them, whereas an assistant must plug deep into various programs, requiring more flexibility and deeper collaboration. With its old-school culture and hardware-style planning process, Apple was out of its depth.

"With Apple, the biggest issue is that they run something like Siri as though it's a hardware thing where up front you can know exactly what you're going to want to do," another former Siri engineer told me. "In reality, you should be humble, you should be trying different things, you should see which ones work, you should invest more in the ones that do work, you should spend more time on those, you should recognize that things will take longer than they seem like they will, because you are doing cutting-edge, machine-learning-type stuff, and you're not really going to be able to predict up front."

In November 2014, when Amazon released the Echo and its embedded Alexa assistant, Apple wasn't exactly surprised by the smart speaker concept. The company had toyed with putting Siri in a speaker in the past, but backed off, perhaps due to quality concerns. The Echo's popularity forced Apple's hand, though, making clear that voice computing would be a new operating layer on top of the internet and apps, challenging the supremacy of screens. So Apple reluctantly joined the fight, and set out to build a speaker with Siri inside.

The HomePod project could've been a turning point for Apple. Unlike with the iPhone, in which Siri was a feature, the HomePod's full experience would depend on the assistant inside. To make the HomePod successful, Apple could've tossed aside its silos, secrecy, and design-led product development process and embraced the Engineer's Mindset, working to incorporate ideas from all levels and divisions. But it again opted for a grand separation over a grand collaboration.

From the outset of the HomePod project, Apple cut off the teams developing it, and some engineers didn't even know what they were working on. "Someone at some point had said, 'It's similar to the Echo,' and that was about as much as I knew," one ex-Apple engineer who worked on the HomePod told me. "A few months before it launched, I happened to be in an office of an engineer who had a cardboard box in the corner of the room. And I was like, 'What's that?' and he goes, 'That's the HomePod,' so I happened to see one that was turned off."

Apple ensconced the HomePod team in a building outside the company's main campus, a facility accessible to only a small fraction of its employees. Nothing had changed from the Siri days. "You don't get the full picture of everything that you're working on," a second engineer who worked on the HomePod told me, adding that he didn't bother going through Apple's disclosure process, preferring to keep all discussion in the room. "The inability to collaborate with certain teams makes the work a bit harder, so you have to find ways to work around that."

Apple's lack of communication tools, meanwhile, created redundant work and unnecessary workarounds, slowing the project down. "Documentation is very lacking," the second HomePod engineer said. "Some of the things you assume there's no documentation on, you'd have to just guess or come up with your own solution, when something already existed elsewhere."

Apple aimed to release the HomePod ahead of the 2017 holiday season. But as the release date approached, some of the device's core use cases were failing. The company then made the rare decision to push the release back. "We can't wait for people to experience HomePod, Apple's breakthrough wireless speaker for the home, but we need a little more time before it's ready for our customers," Apple said in a statement as the holidays approached. "We'll start shipping in the US, UK, and Australia in early 2018."

The extra time did little to mask the device's shortcomings. When the HomePod debuted, even people generally friendly to Apple, like Marques Brownlee, couldn't hide their disappointment. The HomePod is selling so poorly today that eMarketer still lumps its sales numbers in an "Other" category, while the Amazon Echo and Google Home get their own. In 2018, the Amazon Echo had 43.6 million users, Google Home had 19.3 million, and the "Other" category had 7 million.

At the end of our conversation, I asked Siri's founding team member, who has since left Apple, whether he owned a HomePod.

"I own one because I own every single product that has speech in it," he told me.

And his conclusion?

"I think it's very nice with design as usual," he said. "And the assistant itself sucks."

Hands on the Wheel

Imagine, if you will, a beaming Tim Cook taking the stage at Apple's Steve Jobs Theater for the company's biggest announcement since the iPhone.

The theater, a subterranean auditorium on the outskirts of Apple's Cupertino campus, was built for big announcements. And Cook, looking out at the one thousand assembled press, partners, and employees, has brought the goods.

Turning toward the audience, Cook starts off with a nod to the past. "Any company would be fortunate to have one revolutionary product. But here at Apple, we've been lucky enough to build three," he might say. "The Macintosh, the iPod, and the iPhone have all changed lives in profound ways. And today, we're thrilled to show you one more product of this class."

Then, in his cool Alabama delivery, Cook cuts to the chase: "Today, we are introducing the Apple Car," he says. "The Apple Car is a fully electric vehicle with a world-class in-cabin experience that's capable of better than human-level driving. We've built everything ourselves, inside and out, and I promise you will love it."

You could picture the crowd going wild, and Apple certainly has—it's working on this car today. In the mid-2010s, just as the iPhone was nearing "a form that's right," Apple began a quest to build its own autonomous, electric vehicle. Under the code name Project Titan, Apple has dedicated significant resources toward developing this car, believing it can be the company's next "revolutionary" product. But it may be some time before we see Cook give that speech.

The Apple Car—or whatever the company will ultimately call it—is being held back by many of the same factors plaguing the HomePod: Apple has allowed design to dictate to its AI engineers, holding back the technology inside. It's slowed its engineers' progress by keeping them in silos. And its iPhone obsession has kept it from thinking more critically about the right way to build a car. If Apple's HomePod struggles could be cast aside as a one-off, its issues with the car make clear that its problems are systemic.

Inside Apple, the car is viewed as a natural descendant of the iPhone. The iPhone blended world-class hardware (the device) with cutting-edge software (iOS) to set a new standard for phones. This time around, Apple wants to do the same but with a different form of hardware (a car) and totally new software (an autonomous driving system).

"We saw the iPhone as a starting point," an ex-Apple engineer who worked on Project Titan told me. "That was one of the basic problems."

Running the car project as if it were an iPhone refinement, Apple again let design make critical decisions. When you build an autonomous driving vehicle, the software inside is much more important than how the car looks, similar to a smart speaker. But design frustrated its engineers by issuing burdensome top-down edicts instead of listening to what would be best for the project.

Design, for instance, tried to hide the car's sensors, ugly appendages that make your typical self-driving car look like a rolling submarine. But by burying them, design obstructed the sensors' view, limiting the data they could collect and forcing the engineers into suboptimal workarounds.

Design also put its hands on the wheel. After assigning groups to work on the design of the car with and without a steering wheel, design removed the wheel completely, creating further technical challenges for the team now tasked with building for full autonomy. "Some design team said, 'We'll just take it off,'" the ex–Apple engineer said of the steering wheel. "The design team would say, 'Oh yeah, we could get a car without a steering wheel in four or five years.' In reality it doesn't work that way. The lack of a fully supported iterative process is hurting Apple in terms of the new initiatives."

A second ex-Apple engineer who worked on Project Titan also marveled at design's influence. "On top of the engineering challenges, you add these design challenges that make it almost impossible," he said. "Engineers obviously don't have much say on the design. They were forced to work around it."

Apple's silos held the project back further, the second engineer said, telling me the company approached machine learning entirely the wrong way. "Some of us were working on autonomous systems, some of us were working on Face ID. We couldn't talk to each other. We couldn't share what we were doing," he said. "But c'mon, when you walk around, some people are detecting cars, some people are detecting eyes and pupils and facial features. There's lots of things they actually share: they share lots of neural-network models, lots of common practices; I just find it silly. This really slows down artificial intelligence algorithm development."

Finally, Apple management forced its engineers to remain on their lesser internal tech (more on this in a bit).

"Their own stuff is never there; it's always lacking," the second engineer said. "That's the problem for Apple."

In January 2019, Apple moved two hundred employees off its struggling Project Titan. "We have an incredibly talented team working on autonomous systems and associated technologies at Apple," a company spokesperson told CNBC. "As the team focuses their work on several key areas for 2019, some groups are being moved to projects in other parts of the company, where they will support machine learning and other initiatives, across all of Apple."

For Apple, the smart speaker and car projects must feel like a recurring nightmare. The former couldn't hit its target release date and then disappointed. The latter has no release date in sight and is losing staff. Among the two, there's a common line: culture. Secrecy and top-down planning, elements that once bolstered Apple's work, have stymied the company's attempts to invent its way into the future. The absence of the Engineer's Mindset is glaring.

Of the more than two dozen former Apple employees I interviewed for this chapter, many remain loyal Apple shareholders and said they have high hopes for the company's future. Still, in candid moments, their creeping doubt came through. As the first Project Titan engineer told me, "If you couldn't make the speaker smart enough, how could you make the car smart?"

The IS&T Problem

For Apple, making room for employee ideas and finding ways to bring them to life isn't a priority. The company's leadership therefore hasn't placed an emphasis on minimizing execution work with internal technology, as Amazon, Facebook, and Google have. And Apple's internal tools are a source of consternation among its employees.

A group inside Apple called Information Systems & Technology, or IS&T, builds much of the company's internal technology tools—from servers and data infrastructure to retail and corporate sales software—and it's almost universally reviled. IS&T is made up largely of contractors hired by warring consulting companies, and its dysfunction regularly leads to subpar technology. "It's a huge contractor org that handles a crazy amount of infrastructure for the company," one ex-employee who worked closely with IS&T told me. "That whole organization is a *Game of Thrones* nightmare."

Interviews with multiple former IS&T employees and its internal clients paint a picture of a division in turmoil, where infighting regularly prevents the creation of useful software, and whose contract workers are treated as disposable parts.

"There's a cold war going on every single day," Archana Sabapathy, a former IS&T contractor who did two stints in the division, told me. Sabapathy's first stint at IS&T lasted more than three years; the second, only a day. Inside the division, she said, contracting companies such as Wipro, Infosys, and Accenture are constantly fighting to fill roles and win projects, which are handed out largely on the basis of how cheaply they can staff up to meet Apple's needs.

"They're just fighting for these roles," Sabapathy said. "That's all they care about—not the work, not the deliverables, the effort they put in, or even talent. They're not looking for any of those aspects."

IS&T is thus filled with vendor tribalism, in which loyalty to one's contracting company trumps all. "Making a friendship is—like, you wouldn't even think about that," Sabapathy told me, speaking of cross-vendor relationships. "It's not the traditional American way of working anymore. You build relationships when you come to work because you spend most of your time here—that's not there."

Amid the turmoil, internal IS&T clients at Apple can be left reeling as their contractors go dark. "The guy who I was working with got moved to a totally different team and they just replaced him with another guy, and then within a month that guy's gone. And after those people leave, there's a new IS&T project manager and no one told me. I just learned by accident," the ex–Apple employee who compared IS&T to the *Game of Thrones* nightmare told me.

When IS&T's projects are finally completed, they can cause even more headaches for Apple employees, who are left with a mess to clean up. Multiple people told me their Apple colleagues were forced to rewrite code after IS&T-built products showed up broken.

On Quora, a popular question-and-answer site among Silicon Valley types, the question "How is the work culture at the IS&T division of Apple?" has elicited some unbelievable responses. "The engineering quality is extremely lackluster," says the top answer, written by an anonymous user who says they worked at IS&T. "When I first joined, I was absolutely SHOCKED to see how projects were designed and developed. If you compare the code quality to that of a high schooler or a fresh undergraduate, you seriously will not be able to distinguish between the two." I ran this by a former IS&T full-time employee, who said it was accurate.

The next answer on Quora is even darker. "I wanted to share my experience with working in IS&T. Trust me, when I say—this department is worse than most IT sweatshops in India that you have heard of that are a bad place to work for engineers," it reads. "From the day I joined to the day I quit from this department to another, every day was soul sucking and made me curse my life for joining this department."

Sabapathy told me Apple employees' expectations for their IS&T contractors were unrealistic because they only saw the sum total being paid to the consulting companies ($120 to $150 an hour), but the contractors themselves were making much less ($40 to $55 an hour) after the companies took their cut. The approach leaves Apple with lesser contractors but the same high demands, a recipe for disappointment.

When I asked Sabapathy about the Quora posts, she set up the context. "These consultants are from India and they are used to this kind of thing back in India and they are just putting up with the same behavior here too," she said. "It's the same toxic environment we had back in India, which we tried to get away from by coming here. And when we go back into these environments and see the same thing, it hurts."

Apple is not the only tech giant maintaining a large contractor workforce that operates under questionable conditions. Facebook, Google, and Amazon all employ sizable numbers of contractors, with many working as hard as full-time employees but without the same benefits and salaries. These contractor armies are growing fast, and advocates are starting to take note and push for better terms: Google's employee walkout, for instance, made improving contractor treatment core to its protest. Bernie Sanders took on Amazon's lack of transparency regarding its contractors as he pushed the company toward a fifteen-dollar-per-hour wage floor. And in February 2019, *The Verge*'s Casey Newton exposed how Facebook was paying some contract moderators $28,000 per year while paying its full-time employees $240,000 on average (Facebook subsequently raised its moderator wages).

For Apple, fixing its broken IS&T division would not only be the right thing to do from a moral standpoint—it would help the company's business as well. If Apple is going to become inventive again, it will need to give its employees more time to develop new ideas. IS&T could therefore become a division of strength at Apple one day, building tools that minimize execution work while making room for those ideas. But until Apple gives the division a hard look, its employees will be stuck spending their time reworking broken internal software and wishing they were inventing instead.

Face-off

On the morning of December 2, 2015, two terrorists walked into a San Bernardino, California, conference center and started shooting indiscriminately, killing fourteen people by the time they left. The next day—after law enforcement killed the pair—the FBI went to their nearby home and picked up an iPhone 5c.

The iPhone, the FBI believed, could be a key piece of evidence in its investigation into the deceased terrorists and their potential enablers. But there was only one problem: it was locked. The bureau found a four-digit passcode standing between itself and the device's contents. And if it guessed wrong ten times, the phone would erase itself.

The FBI asked Apple to help it unlock the phone. But Apple had no way to bypass the ten-guess limit. Undeterred, the FBI asked Apple to build a back door: a new version of iOS that would allow unlimited passcode guesses. This new iOS, if installed on the iPhone 5c, would give the FBI access to the information it needed. But building it would also make hundreds of millions of Apple devices vulnerable to unwanted access, not just one iPhone. This was untenable to Cook, who rejected the FBI's request. Soon after, the FBI sued Apple to compel them to do it anyway.

The decision to turn down the FBI wasn't easy—what if people were killed as a consequence of information kept secret on that device? Yet Apple was firm in its decision. In a sternly worded letter to Apple customers in February 2016, Cook made clear why he stood on the side of privacy.

"The implications of the government's demands are chilling," he said. "The government could extend this breach of privacy and demand that Apple build surveillance software to intercept your messages, access your health records or financial data, track your location, or even access your phone's microphone or camera without your knowledge. Opposing this order is not something we take lightly. We feel we must speak up in the face of what we see as an overreach by the U.S. government."

The battle positioned Cook as a privacy vigilante, someone who would fight for privacy at all costs, no matter the opponent. Cook made sure to drive the point home. As the fight raged, he appeared on *Time* magazine's cover, where the text "Apple CEO Tim Cook on his fight with the FBI and why he won't back down" overlaid a black-and-white photo of him sitting resolute at his desk.

Never mind that the FBI eventually got into the phone via a third party, and that the legal case was dropped—the showdown was a defining moment for Cook. Apple has always invested in privacy, a result of a business model in which its customers are its users and it doesn't need data-hungry advertisers to foot the bill. But the FBI fight associated the words *privacy* and *Apple* in the heads of anyone paying attention. And Cook has made privacy a core element of Apple's messaging ever since.

That Cook would make privacy so central to what Apple stands for makes sense for a few reasons. First, as phones get to "a form that's right," as Wozniak put it, it's in Apple's interest to make switching off its iOS operating system as difficult as possible. By emphasizing privacy, Cook differentiates Apple's iMessage from Facebook Messenger, Apple's Maps from Google Maps, and Apple's Siri from Google Assistant. Cook and his lieutenants have delivered the privacy message relentlessly at Apple's big events. Stick with Apple's software, they say, and you can feel better about your data. Privacy is now part of Apple's advertising efforts. During the 2019 Consumer Electronics Show in Las Vegas, Apple bought a prominent billboard with the message, "What happens on your iPhone stays on your iPhone."

Amid Apple's privacy campaign, Cook has needled Facebook relentlessly, a logical position given that Facebook owns three massive messaging apps that will soon be interoperable: Messenger, WhatsApp, and Instagram. These apps, like WeChat in China, could make switching off the iPhone easier by supplanting iMessage. Apple hasn't missed a chance to lambast them.

When Facebook was mired in its Cambridge Analytica scandal, Cook made headlines when an interviewer asked what he'd do if he was in Zuckerberg's situation. Apple, Cook replied, would never be in that situation. "If our customer was our product, we could make a ton of money," he said. "We've elected not to do that."

Cook is also running a luxury company whose bestselling product, the iPhone, is reaching parity with the rest of the market. And if Apple can't invent its way out ahead of the field, it will need something to keep the shine on its brand. Something like privacy.

When I watched Brownlee's HomePod video, I wondered if Apple could remain in vogue with YouTubers, our modern trendsetters, if its products weren't as good as its competitors'. And so I called Casey Neistat, a YouTuber and entrepreneur with more than eleven million subscribers, and asked him what he thought.

"Product aside, Apple and Tim Cook have maintained or maybe even accelerated an understanding from my perspective as a consumer that they care about me," Neistat told me. "I trust Apple because of the way they are vocal about privacy. Close my eyes, what do I think of Facebook? I'm scared of Facebook. Every day I debate whether I can afford—because it's my career—but whether or not I can afford to close my Facebook and Instagram account because I'm scared. I have no idea what they're doing with my data, I can't understand it, I don't feel like I have any control. And it's scary."

In my conversations with Apple employees, I learned that the company's commitment to privacy is legitimate. Apple isn't loose with its customers' data as its counterparts can be, sometimes to the detriment of its own products. "Because of the privacy stuff, they don't give the teams access to a bunch of data that the corresponding teams at Google and Amazon do have access to," one of the HomePod engineers told me. "So that really sucks."

In 1997, the year Apple's famous "Think Different" ad campaign came out, Steve Jobs gave an internal talk about the way he sees marketing. "To me, marketing is about values," he said. "This is a very complicated world, it's a very noisy world. And we're not going to get a chance to get people to remember much about us. No company is. And so we have to be really clear on what we want them to know about us. . . . Our customers want to know who is Apple and what is it that we stand for."

In that ad, Apple put forward a defiant message. "Here's to the crazy ones, the rebels, the troublemakers, the ones who see things differently," it said, as footage of Albert Einstein, Martin Luther King Jr., John Lennon, and Mohandas Gandhi rolled. Apple's values were implied: it was among this group, a troublemaker and not a faceless corporation.

Today, Apple is no longer crazy, or a rebel, or a troublemaker. It's a trillion-dollar Goliath with power over the small guys it once counted itself among. Its products, once revolutionary, are now establishment. Its messaging has therefore shifted. What does Apple stand for? The iPhone. And to market it, its value is privacy.

A Drive down 280

As I started wrapping my reporting for this chapter, I wondered where Apple will go now that the iPhone has reached "a form that's right" and the company's inventive muscle seems to have atrophied. So I crossed my fingers and wrote to Steve Wozniak, figuring he might have some ideas. After a few emails, Wozniak told me to meet him the following Wednesday morning at the Original Hick'ry Pit, a barbecue restaurant near Campbell, California, not far from Apple's campus. When the day rolled around, I drove the fifty miles down Interstate 280, which connects San Francisco and Cupertino, and arrived thirty minutes early, wondering if the Apple cofounder would show.

At 10:55 a.m., about five minutes before our scheduled meeting time, Wozniak walked in with his wife, Janet, and business partner Ken Hardesty. Wozniak, evidently a Hick'ry Pit regular, had the staff take us to the back, where we sat and ordered breakfast. As I looked across the table, I saw the man who had partnered with Jobs to bring Apple to life, who'd designed Apple's first computer, and who has remained close to the company since leaving in the 1980s. Wozniak dispensed with the small talk almost immediately, eager to jump in. He urged me to reintroduce the book and start asking questions, and I obliged.

Our conversation began with a discussion of invention, and Wozniak was quick to recall his thoughts on the iPhone. "What's come from Apple? The iPhone," he said. "How much has it even changed in a decade? Not very much. The third-party app store brought all the changes into our life that we often credit to Apple, things like access to Uber."

Apple's inventiveness, Wozniak told me, isn't necessarily about coming up with clever things, but coming up with things that simplify our lives. Throughout our discussion, iPhone refinements came up repeatedly, things like Apple Pay and Touch ID, both of which are delightful features. "We've always been easier to use, simpler, and more to the point, and more like a human being, and not trying to do too much," Wozniak said.

These improvements have helped the iPhone maintain its position at the top of the phone market. And even if people buy the iPhone less frequently, we both agreed, Apple will be fine. "As a user, I'm kind of happy with where things are with Apple right now," Wozniak said. "What if their sales and market share dropped in half? So what. They're still a huge company; it's not going to go away."

But Apple isn't interested in coasting on the iPhone's success. It wants to build a car. It wants HomePod and Siri to succeed. It has bigger plans for the Steve Jobs Theater than showing trailers for shows on Apple TV+, a service meant to make more money from iPhone users ("Because they're in a billion pockets, y'all," as Oprah put it). And it likely wants to do more things we don't have any clue about. To live out these dreams, Apple will need a culture change.

After talking over the Engineer's Mindset with Wozniak, I asked him how Apple could be more inventive. The Apple cofounder initially tossed aside the question, telling me he didn't know if Apple could be "more" inventive because he wasn't inside the operation.

But then, soon before our meeting concluded, Wozniak answered the question. "Let the lower-level managers make the decisions," he said. "More responsibility to the lower levels."

CHAPTER 5

SATYA NADELLA AND THE MICROSOFT CASE STUDY

When Microsoft acquired the advertising firm aQuantive for $6.3 billion in 2007, the mood inside the just-purchased company wasn't exactly festive.

"I'm not working for fucking Microsoft," one employee said upon hearing the news. A day later, he quit.

The somber scene was atypical for a startup just acquired for big money. Employees almost always celebrate such events, understanding that the cash, stability, and support involved will free them from the stress of startup life and allow them to focus on their work. But with Microsoft buying, this wasn't the case.

aQuantive became the world's most valuable ad-tech firm by embracing the Engineer's Mindset. Ideas flowed freely inside the company. Managers removed red tape. And employees invented with abandon. "You could walk into any VP's office, you could have a conversation with folks, and there wasn't a lot of internal competition,"

Abdellah Elamiri, an aQuantive employee at the time of the acquisition, told me. "Teams were free to release whenever they wanted, and they had a lot of autonomy."

Microsoft was different. Under its then-CEO Steve Ballmer, who rose through the sales function, the company was bureaucratic and slow, and clung to the past. Focused on protecting its lucrative legacy businesses, Windows and Office, Microsoft prioritized profit over invention, developing a command-and-control culture that optimized for the short term. The alpha males running Windows—the dominant desktop operating system in the age of personal computers—almost always got their way.

"Microsoft was an old, hard-core, smartest-guy-in-the-room culture," Robbie Bach, the company's former president of entertainment and devices, told me. "It was definitely a place where you better stand up, have an opinion, and be strong about it."

Coming into Microsoft, aQuantive's employees knew they were in for a culture clash. And after a brief honeymoon, "the edicts started coming," Elamiri said. In one episode, the Windows team almost killed aQuantive's core business by deciding to ban the cookies in Internet Explorer that underpinned its ad targeting. They called off the move only after Brian McAndrews, the aQuantive CEO who remained post-acquisition, found out secondhand and fought it fiercely.

Microsoft's substandard internal technology caused further frustration among aQuantive's employees. AI was still in a deep freeze, and the company was so dedicated to Windows it refused to use tools other companies built. When Microsoft employees brought Apple products into work, their colleagues ostracized them, even if they were developing for these devices. Ballmer, who once pretended to smash an iPhone in a meeting, set the tone. "One of the initial problems was that if it wasn't a Microsoft technology, they wouldn't bring it on board," Elamiri said. "If it's not built in Redmond, we don't do it."

By 2012, Microsoft wrote down aQuantive's $6.3 billion value to virtually zero. And it was clear to everyone involved that culture was the culprit. "No amount of explanation about ad revenue versus software revenue or Google's plan to make software free could refocus a Windows-obsessed culture," a former aQuantive manager told *GeekWire* at the time.

The same week Microsoft wrote down aQuantive, *Vanity Fair* circulated an advance copy of an article that would brand the Ballmer years "Microsoft's Lost Decade." The article made clear that the aQuantive debacle was no anomaly. "What began as a lean competition machine led by young visionaries of unparalleled talent has mutated into something bloated and bureaucracy-laden, with an internal culture that unintentionally rewards managers who strangle innovative ideas," the article said.

As aQuantive faded, Elamiri transferred to Microsoft's Skype group. And from there, he saw the winds of change blow in. Ballmer stepped down in 2014, giving way to Satya Nadella, a twenty-two-year Microsoft veteran.

A "consummate insider," in his own words, Nadella understood Microsoft needed to reinvent itself to survive—or *Hit Refresh*, as he titled his bestselling book. The company's Windows obsession caused it to miss the mobile revolution. Its rivals, Apple and Google, now owned the world's most important operating systems. Tightly holding on to Windows would no longer be tenable. Microsoft needed to risk its core business and focus on its remaining bright opportunity—cloud computing—or endure "irrelevance, followed by excruciating, painful decline, followed by death." So Nadella took a lesson from his neighbor across Lake Washington and returned the company to Day One.

To reinvent Microsoft, Nadella had to reimagine its culture first. The company had too many barriers preventing ideas from traveling across its divisions and had lost its inventive muscle. In an attempt to reverse this, Nadella made Microsoft look a lot like pre-acquisition aQuantive. He took a hammer to the company's hierarchical structure and sent the alpha males packing. He sparked invention, using AI to cut down execution work. And he facilitated collaboration by tearing down silos, emphasizing empathy, and breaking apart the all-powerful Windows group.

"Microsoft's lost decade could serve as a business-school case study on the pitfalls of success," the *Vanity Fair* article said.

Today, it's time for a new case study. Microsoft has rebounded historically under Nadella, who's led its revival by prioritizing the future over its past and embracing the Engineer's Mindset.

Nadella's Day One

On an uncommon rainy day in Palo Alto, California, Professor Susan Athey welcomed me into her third-floor office at Stanford University's Graduate School of Business. Athey is a rare straight-talker in Silicon Valley, a no-nonsense academic who once helped me debunk a study that ignored the law of supply and demand. She had also served as Microsoft's chief economist under Steve Ballmer, which made her particularly well suited to discuss the company's "lost decade" and how it climbed out.

I caught Athey in the middle of a busy day packed with back-to-back meetings. Her office on Stanford's campus was brimming with books piled shoulder high and a whiteboard filled with markings. As I took my seat, Athey leaned back in her chair and began to tell the story of how Microsoft's past held back its future.

At Ballmer's Microsoft, she said, there were two divergent factions. One faction—which I'll call the "asset milkers"—believed Microsoft should milk its lucrative Windows business for all it was worth. The other faction—the "future staters"—believed Microsoft should risk cannibalizing Windows to build for the next state of computing.

"Some felt—and there's a rational view—that you had this great asset, and you should just get as much out of it as you can on the way down until it dies," Athey said, referring to Windows, which for over a decade owned more than 90 percent of the desktop operating system market. "The second view is no, we think we actually can be successful and profitable in the new state, but that's going to require not milking everything you can out of the old state."

The critical battle between the asset milkers and future staters was over the cloud. In the early 2000s, Microsoft had a division called Server & Tools that helped its customers build programs they'd install and run on desktop computers. By 2008, Server & Tools was a $13 billion business that had turned in twenty-four consecutive quarters of double-digit growth, making up 20 percent of Microsoft's total revenue. Some Server & Tools customers created programs they sold to others, and many developed applications for internal use. When internet speeds grew faster, companies started to host internal applications (such as email servers) externally and build software for use on the web browser as opposed to desktop (aka cloud computing). Seeing this early move to the cloud, Microsoft had to decide whether to support it and to what extent.

Cloud computing, though promising, was a threat to Microsoft's Windows business. If software went to the cloud, people wouldn't *need* Windows. They'd be able to access applications on any operating system, whether it was Windows, Apple's macOS, or Google's ChromeOS. And they wouldn't need Microsoft's pricey internal servers. For the asset milkers, pivoting the lucrative Server & Tools division while undermining Windows would be disastrous. For the future staters, such a pivot would allow Microsoft to build an early lead in cloud services, which could become a significant business.

In their push for the cloud, the future staters ran into a roadblock: Microsoft's own customers, who told the company they'd never move to the cloud. These customers, generally chief information officers (CIOs), bought software for all departments, installed it, secured it, maintained it, and evaluated it. The CIOs wanted no part in a future where individual departments like sales and marketing could subscribe to software hosted on the web, minimizing their power and influence. "If you go to this guy and you say, 'Would you like to shut down your operation and send it to the cloud?' you get a pretty universal no," Athey said.

Microsoft listened to these CIOs for a time. But when the company's corporate strategy team and Athey ran a deeper analysis, their findings ran counter to what they were hearing. "Over a period of years, all those CIOs would either be moving to the cloud or they would be fired," Athey said of the results. While Microsoft waited, Amazon built AWS and took the lead in cloud services. By 2013, the year Ballmer announced he was stepping down, AWS controlled 37 percent of the $9 billion "infrastructure as a service" market and was growing 60 percent annually. Microsoft was far behind, with 11 percent of the market.

Microsoft faced a similar decision with Office. The Office suite was a main draw for Windows devices, which many people bought to use Word and Excel. Making it available across mobile devices and web browsers threatened Windows. Putting Office on the browser could also cannibalize its own healthy desktop sales. The asset milkers wanted Office to be available primarily via desktop installs. The future staters, looking ahead to the coming age of mobile and cloud computing, wanted it everywhere.

Microsoft's strategy for Office during the Ballmer years generally followed the asset milkers' desires. Instead of building Office for web when Google released Docs and Sheets, Microsoft kept Internet Explorer slow and held Office off-line. Some years later, Microsoft put a limited version of Office on the web and released it for mobile—but only on Windows devices. Even then, Microsoft kept Office's web version so quiet that its employees didn't even know it was live.

"One thing that drove me crazy," Athey said, "is that I remember going around Microsoft when the first web versions were available and I would give presentations and I would pull up web Office, and people were like, 'I didn't even know that existed.' It was not uncommon that people didn't know it existed inside of Microsoft. And outside, people were like, 'Really, there's a web Word?' "

In the heat of these battles, Ballmer promoted Satya Nadella, the executive in charge of Microsoft's Bing search engine, as head of Server & Tools. Nadella was not like the rest of Microsoft's top executives. He didn't have an outsize ego. He didn't shout his opinions. He stayed out of Microsoft's political infighting—where constant battles between the asset milkers and future staters, and just about everyone else, was standard. And having seen the future of computing while working on Bing, Nadella didn't view the company's existing products as sacred either.

Though Bing is still the butt of jokes—"Cue the evil laughter and organ music," the *Vanity Fair* article said, introducing it—Nadella's experience working on it taught him a lesson in the importance of cloud and AI. A search engine is a powerful program built for use on a browser. When you build one, you're building on the cloud. Bing, like Google, sorts through a massive amount of data (nearly all the websites on the internet, the content inside them, the links pointing between them) and tries to make sense of it—a task particularly well suited for machine learning. When Nadella was senior vice president of Microsoft's Online Services division in the late 2000s, a group that included Bing, he got a crash course in the future of the internet.

"In running the search business, you needed to understand all the costs of the data centers, the efficiencies involved. You had to be an expert in deployment in the cloud," Athey said. "You also had to be an expert in A/B testing platforms, and continuous improvement, and machine learning. Satya was an expert in all of those things."

When Nadella took over Server & Tools in 2011, he understood that simply providing servers and tools to companies building software for desktop machines wasn't going to be viable.

Watching Amazon Web Services' early success, and reading his economics team's analysis, Nadella decided that acting any slower would set Microsoft back. While working on Bing, Nadella learned how difficult it was to be a distant number-two player in a market, and he wasn't eager to repeat the scenario. Despite the risks to the core Windows asset—not to mention a thriving Server & Tools business—Nadella made clear that Server & Tools under him would focus on enabling cloud computing. It was future state or bust.

"It was a bit terrifying he was persuaded by the analysis," Athey said.

By late 2013, Ballmer, who did not respond to an interview request, understood he'd outlived his usefulness for Microsoft. The asset milkers he empowered lost their credibility as mobile and cloud rose to dominate the technology landscape. His appointment of Nadella was indeed the turning point for the company, Athey said. But for Ballmer, it came too late. In August of that year, he said he'd step down.

Ballmer left Microsoft in a tough but manageable position. His final major act, a $7.2 billion acquisition of Nokia that Microsoft also wrote down to zero, left a lasting impression of incompetence. But inside the company's Server & Tools division, Nadella was building Microsoft's future, one that said no to Windows orthodoxy and yes to the cloud and mobile—the future staters' dream. On February 4, 2014, after a quick search, Microsoft named Nadella CEO.

Democratic Invention

When Nadella took control of Microsoft, there was little question about the strategy he'd pursue. The new CEO's track record on Azure and Bing made it clear he'd orient the company around a mobile-first and cloud-first vision. In an email to employees his first day on the job, he drove that point home.

"Our industry does not respect tradition—it only respects innovation," Nadella wrote. "Our job is to ensure that Microsoft thrives in a mobile and cloud-first world."

For Nadella, strategy was the straightforward part of the job. The thorny area would be culture. The Microsoft he inherited was more interested in refining Windows and Office than in building new things, making it an unfriendly place for employees with big, new ideas. The company's leaders, accustomed to monopoly life, also assumed people would buy its products simply because they were coming from Microsoft, causing them to lose touch with what it took to build products people wanted. As Microsoft entered the new, competitive cloud-services market, that mentality wasn't going to cut it.

"Microsoft typically did not care about the user," one former product manager told me. "The mentality of most of the product groups was, 'We'll build it, and they'll come; don't worry about it.'"

To spark a new era of invention inside Microsoft, Nadella first gave his employees permission to come up with big ideas again. He set the tone right in that first-day email. "We sometimes underestimate what we each can do to make things happen," he wrote. "We must change this."

Nadella then exposed his leadership team to as much startup thinking as possible, bringing the founders of companies Microsoft acquired to its annual leadership retreat and inviting startups into Microsoft's Redmond, Washington, campus to teach his leaders to think like early-stage companies. "Different startups came in and talked to us about their businesses, and their culture, and how they ran their companies, just to expose us to different thinking and new ideas," Julie Larson-Green, a twenty-four-year Microsoft veteran who rose to a chief experience officer role before leaving in late 2017, told me.

Nadella also expanded Microsoft Garage, a physical and virtual space for product experimentation, creating a public website where Microsoft would release experimental apps. The website's copy today reads distinctly Amazonian. "Our motto 'doers, not talkers' continues to be the core of what we are," it says.*

And in a recurring series in his Friday staff meetings called "Researcher of the Amazing," Nadella started having employees from across the company who built inventive new programs come in and present.

To make Microsoft's new inventive energy useful, Nadella needed to channel it into building things people wanted. So he instructed his product teams to investigate what their customers were

---

*Microsoft Garage's "About" page once included a more direct homage to Amazon's leadership principles. "The Garage has a bias for action," the page said in September 2019. Almost immediately after I brought this up to Microsoft on a fact-checking call, the sentence disappeared from the website. A Microsoft spokesperson called it a coincidence.

experiencing in their real lives, focusing on their needs first instead of Microsoft's. Build with empathy, he told them.

"It's not just thinking about what the customer wants, but being the customer," one current Microsoft product marketing manager told me.

"The philosophy change was starting to move away from talking about the product and capabilities, and toward talking about who exactly is going to use this, and why, and how are we going to differentiate," Preeta Willemann, an ex–Microsoft product manager who worked on a presentation program called Sway, told me.

Willemann said that about a year after Nadella came on board, her entire team—"Product managers, design, engineers, everyone"—put its work on hold for two weeks to brainstorm the types of people who might want to use their software. Then, they interviewed those people to see what their lives were like. "We were just trying to start from, who are you and what opportunities may there be in your daily life, not thinking about our software at all," she said. "Once we had identified those opportunities, we worked with them to see if our software could address those specific opportunities."

After holding these meetings, the team realized that it appreciated some product features more than its customers did. Microsoft was building a product with lots of snazzy features, such as 3-D visualizations, but the product's target customers, small businesses, wanted something more straightforward. "For the most part, they were largely uninterested in the software we were building," Willemann said. The team then adjusted according to the feedback. "It was incredibly clarifying," she said.

Building with empathy was particularly useful for Microsoft's cloud offering, now called Azure, which Nadella had to sell to customers who said they didn't want it. Nadella, a cloud customer himself when he worked on Bing, had the Azure team put itself in its CIO customers' shoes. For these customers—banks and other large, slow-moving companies—moving to the cloud would take many years. So Microsoft built for their use case, providing hybrid services composed of both cloud and desktop support, a move that kept the CIOs relevant while gradually moving their companies into the future. This model differentiated Microsoft from Amazon Web Services, which generally sold to companies that built their entire software applications on the cloud, per Microsoft's internal research.

"Microsoft has been an enterprise vendor for a long time. CIOs trusted Microsoft, and they still trust Microsoft," Sid Parakh, a portfolio manager at Becker Capital Management, which is betting big on Microsoft, told me. "When Microsoft had a good product to offer, their customers were willing to latch on to it."

Nadella then had to create more time for his employees to come up with ideas, and to find ways to channel them to the right people. For that, he leaned on AI.

In Microsoft's sales organization, as in most sales organizations today, the company's sales reps spent much of their time digging through customer relationship management (CRM) tools trying to figure out whom to call, what to tell them, and which calls to prioritize. This work adds little value and can be minimized with machine-learning technology, which can sift through sales data and predict which deals are most likely to close by examining what's worked in the past for similar customers.

Applying machine learning to sales should've been an obvious move for a company like Microsoft, which has some of the world's top AI talent. But it wasn't seriously considered until Nadella reorganized the company's AI division in 2016 and instructed a part of it to focus on more practical applications. "We are infusing AI into everything we deliver across our computing platforms and experiences," Nadella said at the time.

Following the reorganization, the company set up a venture capital–style pitch committee for its AI researchers. If the committee liked a researcher's pitch, it would give them some resources and a few weeks to build their prototypes. Then, if the researchers hit certain milestones, they got a few more months to build the product.

At the time, a researcher named Prabhdeep Singh was thinking of leaving the company and starting out on his own. Upon hearing of Singh's plans, an executive in the research organization advised him to hone his skills by pitching the committee, and then make his move. Singh decided to go along.

Thinking about the ways Microsoft could apply machine learning, Singh saw the opportunity in sales most clearly. "If you have to use artificial intelligence, the place where you can see the results instantly and the quickest were in sales and marketing," he told me, "because if it works, you would see an uptick in revenue right there."

Singh pitched the committee and got approved to build a feature called the Daily Recommender, then code-named Deep CRM. Using machine learning, the Recommender sorted through all possible actions a Microsoft sales rep could take, and suggested the most valuable actions, one by one, giving the rep the option to accept or skip these actions. The tool cut away the drudgery of combing through the dreaded CRM (and other systems) to figure out what to do next.

The Daily Recommender, still live today, considers one thousand data points per customer to generate its suggestions. This includes what happened in similar situations with other accounts, even those accounts not assigned to the reps themselves. It can recommend actions like calling X account because they just got funding and are growing, or calling Y account because their usage of a product is dropping and they are likely to churn.

"It's taking the opportunities and weighting them so that the most likely ones are at the top," Norm Judah, Microsoft's former enterprise CTO who oversaw the feature, told me.

The Daily Recommender learns as it goes. If a salesperson makes it through fifty recommendations in a day, it adapts and gives them more. If a rep makes it through twenty, it learns to give them less. If a rep closes a deal after taking action on a recommendation, the system learns it was probably a good recommendation. If a rep skips a recommendation and closes anyway, it learns its recommendation was off.

"The seller has an understanding of the cultural behavior of the customer, or the sequence that they buy things in," Judah said. "But as the system builds more and more history, that intuition actually becomes algorithmic."

The Daily Recommender is mostly used by Microsoft's small and midsize business salespeople. For larger businesses, the company uses other machine-learning tools to suggest the next product its customers are likely to buy. When Singh's team started to introduce these systems, he worried they'd generate a backlash among Microsoft salespeople who'd feel they'd be more productive without them. But not long into the experiment, the reps who didn't have them—the control group—began demanding them.

Putting AI into sales, Singh said, helped generate $200 million in extra revenue by the time he left Microsoft. More important, it helped sales reps spend less time on execution work.

As these machine-learning systems cut down execution work, they freed Microsoft's sales team to spend more time speaking to customers. And as Microsoft moved to a more empathetic product development process, these conversations—held by the people in its organization with the most profound relationships with its customers—influenced the product direction.

"Pretty much everything was self-service on the sales side," Singh said. "The salespeople were clarifying the needs of the customers, quantifying those needs, telling them how Microsoft products could satisfy those needs, and then bringing the feedback from the customer to the product teams."

Microsoft then employed a software tool called OneList that centralizes feature and capability requests, using it to channel product ideas from sales to its product teams. "All of that stuff gets aggregated back into one place, and then engineering leadership is held accountable back to that list," Judah said. "The identification is one thing, but having a process that takes the identification and puts it back into a plan, or prioritization, that's the important part."

Today, Microsoft has a version of the Daily Recommender in Microsoft Dynamics, its cloud-based customer relationship management system, where it's known as the Relationship Assistant. Prabhdeep Singh, meanwhile, moved to UiPath in 2018, where he's ensuring his practical-minded application of AI will be available beyond Microsoft's orbit. And Microsoft is building products people want to use again.

Over lunch at the Diner of Los Gatos, an old-fashioned Silicon Valley eatery, Microsoft chief technology officer Kevin Scott told me machine learning–powered systems like the Daily Recommender are present across Microsoft.

"I'm seeing lawyers, I'm seeing HR people, I'm seeing folks in finance who are all using these tools to solve problems," he said.

The company's most promising tools are making it possible for anyone to invent, and Scott started describing them. For instance, Lobe, a company Microsoft acquired in 2018, enables people with limited technical skills to build machine learning–powered programs. One of Lobe's cofounders—with little knowledge of the underlying AI—used it to build a program that monitors the water tank levels in his off-the-grid house. With a web camera and some labeling, Scott told me, Lobe was able to identify a weight on the tank connected via rope to a float inside. As the weight moved up the tank, the program understood the water was going down, and updated the tank's levels.

"You feed these images into a machine-learning system. And you say go and it builds the model," Scott said. "That's crazy powerful."

Microsoft's tools are also minimizing execution work for coders themselves. A program called Visual Studio Code uses machine learning to predict engineers' code as they write. "It looks at the context of what you're typing, and what it knows about the structure of your code, and the programming language, and it suggests to you what you might want to type next," Scott said.

Microsoft's internal technology—much of which the company

builds internally and then licenses for external use—can cut down execution work in companies across the economy, potentially helping them become more inventive. "What gets me out of bed in the morning is this idea that we have a responsibility right now to put these tools into the hands of as many people as humanly possible. To empower people to create with advanced machine learning and AI, at a massive, massive scale," Scott said.

As we awaited the check, I asked Scott what he makes of the argument that invention will be limited to a small segment of people, mostly programmers.

"Insane," he said.

Constraint-free Hierarchy

For Nadella, sparking inventive ideas would be useful only if his executives were receptive. And as he reworked Microsoft's culture, he had to get them to start listening.

Ballmer's Microsoft didn't value ideas from the rank and file. It was obsessed with refining its core products and had little use for ingenuity. No simple channel existed for the rank and file to bring ideas to the top. Employees were discouraged from speaking with people above their bosses, unless their bosses were present. And meetings became exercises in shouting over listening.

After witnessing Microsoft's hierarchy constrain people and ideas for years, Nadella addressed his frustration with it in *Hit Refresh*. "Our culture had been rigid," he wrote. "Hierarchy and pecking order had taken control, and spontaneity and creativity suffered as a result."

To free people and ideas from Microsoft's hierarchy, Nadella employed tactics straight out of the Facebook playbook. He built a culture of feedback, having his employees meet with their managers every quarter for feedback sessions called "Connects." He started hosting Q&As with employees. And he went out and listened.

"Over the first several months of my tenure, I devoted a lot of time to listening," Nadella wrote. "Listening was the most important thing I accomplished each day because it would build the foundation of my leadership for years to come."

This listening campaign was a departure from Ballmer's style, but consistent with Nadella's conduct. From his early days at the company, he'd take young employees out for meals, simply to hear their thoughts on where the technology world was going. "Satya would seek my opinion out," one former Microsoft employee who worked with Nadella in the early 2000s told me. "It was unthinkable that some senior exec would ask some random twenty-three-year-old program manager, 'Hey, how is this startup doing?' No other exec would give me the time of day."

Nadella's style made the company's leadership newly approachable, several current and former Microsoft employees told me. "In every meeting, every situation, he would be very open about what he knew and didn't know," Julie Larson-Green, the former chief experience officer, said. "That made it okay for other people to talk about how they felt."

Nadella also quashed a relic of Microsoft's hierarchical culture, one you can still watch on YouTube today. Under Ballmer, the company gathered its employees together for an annual meeting where he would flail around onstage, shrieking and screaming catchphrases like "I love this company" while music blasted. Videos of these events have racked up millions of views online, appreciated primarily due to their comedic value. Inside Microsoft, rumor has it that Ballmer would chug a full honey bear before these appearances. On YouTube, commenters tend to think it was another substance.

Ballmer's antics make for a fun watch, but they epitomized Microsoft's hierarchical nature under his rule, where executives tended to shout orders at employees instead of listening to them. After taking over as CEO, Nadella put an end to this annual meeting. In place of the lights and music, he introduced the "One Week," a yearly employee gathering that put a hackathon, not a CEO-led pep rally, at the center.

"We had administrative assistants, and people in the legal department, and our finance department that would help give ideas," Larson-Green said of the hackathon. "We build products for regular people to improve their everyday lives. You need to really live that life and understand those people and what they care about."

Nadella also tackled Microsoft's glut of middle management. When ticking down the list of issues responsible for the company's failures, the *Vanity Fair* article paid this class of people special attention. "More employees seeking management slots led to more managers, more managers led to more meetings, more meetings led to more memos, and more red tape led to less innovation," the article said. "Everything, one executive said, advanced at a snail's pace."

Nadella employed a clever strategy to get Microsoft's middle management to buy into a system that minimized their gatekeeper power: he squeezed them. In leadership meetings, Nadella would hammer home the importance of not being a bottleneck. And by writing *Hit Refresh*—and giving a special, annotated edition to the Microsoft rank and file—he instilled his vision in his employees, who'd preach his anti-hierarchy gospel to their managers too. "Satya's pushing down, and they're pushing up on this culture change," Larson-Green said. "That's why he didn't have to really replace the middle layer as much as grow the bottom layer that needed to be there."

Microsoft today is still filled with relics of the Ballmer-age hierarchy. Employees still complain of bad managers and roadblocks. But under Nadella, ideas flow upward in a way they didn't when his predecessor ruled. "People were more curious, more eager to learn, more interested in what customers thought," Larson-Green said. "They felt less pressure to be the one with the answers and more pressure to understand the problems."

Collaboration

Microsoft is a company of conflicting interests. When you stack its teams' ambitions against each other, they're often in direct opposition.

Microsoft Office, for instance, is in natural conflict with Microsoft's devices. Office wants to be available everywhere so it can reach the broadest possible market. But Microsoft devices' interest is to keep Office exclusive to its own products to make them must-buys for Word and Excel fanatics. Windows and Azure are in similar conflict: Azure wins add up to Windows losses. Conflicts of this nature exist down the chain at Microsoft. And with the help of poor management practices, they've led to legendary infighting.

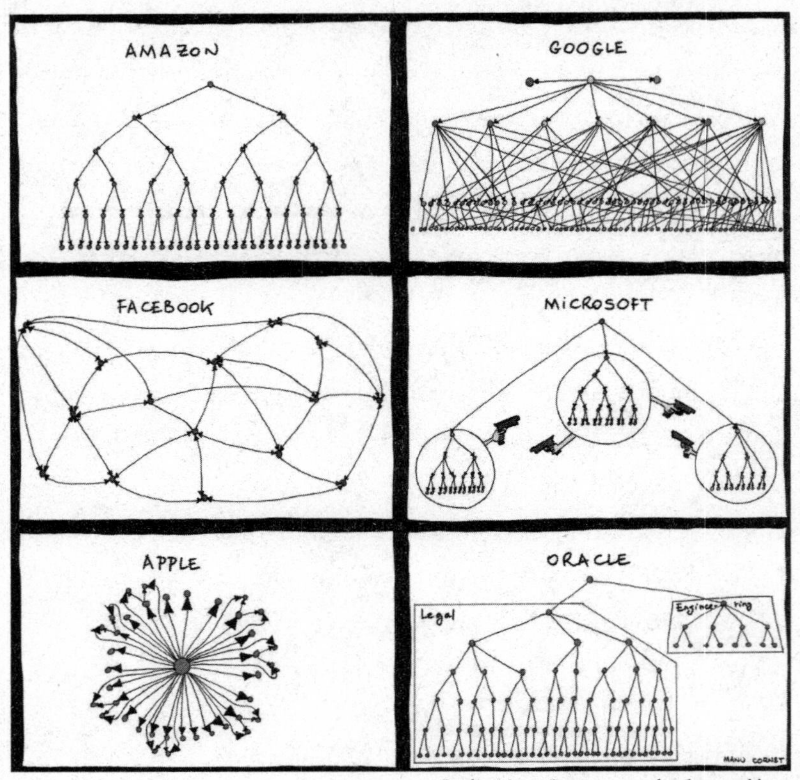

Credit: Manu Cornet, www.bonkersworld.net

When Nadella took over Microsoft, getting the company's divisions working together was a particularly tricky challenge. For Microsoft to move into the future, it couldn't have its employees sabotaging one another. Nadella had to teach a company of infighters to collaborate to see his vision through.

"We are one company, one Microsoft—not a confederation of fiefdoms," Nadella wrote in *Hit Refresh*. "Innovation and competition don't respect our silos, our org boundaries, so we have to learn to transcend those barriers."

To reignite collaboration inside Microsoft, Nadella told his employees to come to work with a "growth mindset," using a concept developed by Stanford psychologist Carol Dweck. In *Mindset*, her 2007 book, Dweck argued that individuals who believe they can grow are more likely to achieve than people with a "fixed mindset," who believe they have a natural ceiling. Nadella ran with the idea and applied it to his company. For Microsoft, employing a growth mindset would mean focusing on what could help the company grow the most, thinking beyond individual divisions and their limitations.

"We need to be open to the ideas of others, where the success of others does not diminish our own," Nadella wrote in a 2015 email to all employees.

Soon after that email, stickers urging Microsoft employees to employ a growth mindset began appearing in Microsoft's conference rooms. And employees began reinforcing the message to one another. "Growth mindset, that's a really big phrase; they just keep on repeating it," one senior product executive at Microsoft told me. "They talk about it on the intranet, in the company all hands, in the division all hands, in performance reviews. It's everywhere. You can't escape it."

Operating with a growth mindset would mean freeing Office to run on all operating systems, forgoing Microsoft devices' specific interests in service of more significant revenue potential. Nadella drove this point home by demoing Office for iOS devices in his first public product presentation. Soon, Apple devices began appearing on Microsoft's Redmond campus.

"We don't care what you're running anymore. We just care that you're buying our services—Office, Dynamics, Azure, all of those—cross-platform," Stephan Smith, a former Microsoft senior consultant, told me. "I'm telling you, this is why Microsoft is going through the roof. They took off the restraints."

After seeding the growth mindset, Nadella reworked Microsoft's structure to support it. On March 29, 2018, he oversaw Microsoft's "Biggest Reorganization in Years," according to *Bloomberg*, splitting the Windows division in two. The bulk of Windows went to a new Cloud & AI division, where it would pair with Azure, its former archnemesis. The Windows devices team went to a new Experiences & Devices division, where it would be paired with Office and have to work out its misaligned interests. The name Experiences & Devices was no accident; it was about the experience first, then the devices.

"We can't let any organizational boundaries get in the way of innovation for our customers," Nadella said in an email announcing the changes. "This is why a growth mindset culture matters."

Nadella also reimagined the way Microsoft handled acquisitions. When Microsoft acquired LinkedIn for $26 billion in 2016, Nadella put LinkedIn CEO Jeff Weiner in charge of integrating the two companies. To ensure this went smoothly, Nadella added Weiner to his senior leadership team and made him a direct report, signaling to LinkedIn's employees that their ideas would be taken into account.

"What it said to the LinkedIn employees is, 'Okay, you can relax, the boss that you have had for forever and a day, who you already trust, is not going to do anything silly or bad,'" Kevin Scott told me. "And the thing it communicated on the Microsoft side is that Satya was dead serious about having LinkedIn have the level of autonomy that LinkedIn thought that it needed."

Microsoft's LinkedIn acquisition was a marked departure from how it handled aQuantive, and it's led to results: LinkedIn's revenue is rising by 25 percent annually.

The final element of Nadella's plan to inspire collaboration at Microsoft was to change the way the company evaluated employees. Microsoft had a long history of performance evaluations called "stack rankings" that pitted its employees against one another. The dreaded system forced managers to rate their reports on a bell curve. No matter how good a team was, or how evenly talented its members were, a preordained number of people would get great reviews, and a set amount would get poor reviews.

"Let's say your whole team was the same skill level; you still had to force it," one former senior manager told me. "Somebody was going to get a huge bonus, and someone's going to get borderline fired. Not that extreme, but it's going to happen."

Because of this, employees sabotaged one another. And the company's most talented people went out of their way not to work with one another. "Microsoft superstars did everything they could to avoid working alongside other top-notch developers, out of fear that they would be hurt in the rankings," the *Vanity Fair* article said. "Microsoft employees not only tried to do a good job but also worked hard to make sure their colleagues did not."

Before Ballmer left the company, he dismantled stack rankings. With a fresh slate, Nadella developed a system radically different from his predecessor's. Individual impact at Microsoft today only counts for one-third of a performance evaluation. The rest of an assessment looks at what someone's done to help others succeed, and what they've done that builds on others' work. There is no forced ranking.

"The 'how' became as important as what got delivered," Larson-Green told me. "If you're shutting other people down in meetings, or you're not collaborative, or you're a jerk, you aren't going to get as rewarded as someone who had an equal contribution but did it in a way that made the team stronger."

Though Nadella's making progress, Microsoft is still no perfect place to work. Treatment of women in the workplace is one particular area of weakness. In spring 2019, an email chain went around Microsoft's office in which female Microsoft employees vented about how they were mistreated in Nadella's Microsoft. One woman said she was only given menial tasks despite being in a technical role, another said she was asked to sit on a high-ranking coworker's lap, and another recounted being called a "bitch" and said it was pervasive throughout the company. It didn't help matters that Nadella once said women shouldn't ask for raises, and should have faith "the system will actually give you the right raises as you go along," a comment he's since apologized for. Nadella was on the email chain, according to *Quartz*'s Dave Gershgorn, but Microsoft's head of human resources was the one who responded. Nadella wrote the entire company with a subsequent email, according to a Microsoft spokesperson, who declined to share the email.

At Microsoft, this type of demeaning language was no anomaly. "People had come forward and told me and others that an engineering colleague had made racist comments, sexist comments, and is a bully," one former Microsoft manager told me. "When I called it out in his performance review, I got told that this guy is too valuable to the company, he knows so much in a field that very few people understand, and that it would just be super hard for us to lose someone like that. And I'm just like, 'Holy shit.'"

Nadella's changes, though still far from complete, have turned Microsoft into a better place overall. And he's won a fan in Abdellah Elamiri, who watched Microsoft go through real change in his ten years at the company. "Microsoft was evolving to give more autonomy and move away from command and control," he said. "It's not all about Windows. It's a matter of doing the right thing for the business and the customers."

Microsoft's New Decade

In August 2019, seven years after *Vanity Fair* published "Microsoft's Lost Decade," I called up Kurt Eichenwald, the man who wrote the story.

In the years since Eichenwald's article came out, Microsoft has transformed. The company is no utopia—current and former employees still complained to me about bad managers, silos, egos, and obstructionism—but it's in a different place today than it was in July 2012. I wondered if Eichenwald was surprised.

After a few rings, he picked up the phone. Soon into our call, he pointed to the different reactions he'd received after publication. Microsoft's very top layers hated it, he told me, while those in the middle and a little higher called to thank him. "What that said to me was, you had a very serious disconnect between the senior management and the actual operations of the company," he said.

"Culture," Eichenwald told me, "is the single most important part of having an effective corporation. When you have policies driving culture in a direction that senior management doesn't recognize, but are causing a lot of difficulty, the senior management either will change dramatically or will ultimately be decapitated. Because it's not sustainable."

The top of Microsoft did change. And Nadella, someone who spent years in the middle of the company, has taken it in a new direction. "When you're playing on the field, it's hard not to see how the plays are going," Eichenwald said. "You're out there actually working with the consequences of poorly thought-out decisions. You're going to recognize things for what they are."

Recognizing things for what they were, Nadella unhooked Microsoft from its Windows-centric thinking and had the company reinvent itself before "the Asset" was fully drawn down. He did this by running Microsoft with the Engineer's Mindset, democratizing invention in the spirit of Amazon, freeing people and ideas from hierarchy in the style of Facebook, and doing the hard work to inspire collaboration in the mode of Google. Using internal technology to cut execution work, Nadella made it possible for Microsoft to turn around before the competition kneecapped it.

Nadella's cultural change has led to real business results. Microsoft's market cap listed in the *Vanity Fair* article was $249 billion. Today the company's market cap is more than $1 trillion. Office and Azure are selling better than ever, and Windows is holding steady.

"Lots of companies can play Lazarus," Eichenwald said, "if they learn the lessons from their failures."

CHAPTER 6

A LOOK INTO THE BLACK MIRROR

The science fiction series *Black Mirror* debuted to little fanfare in 2011. At the time, society felt mostly positive about technology. And the series, which imagined modern technological progress taken to its most dystopian ends, faced an uphill battle.

Once *Black Mirror* hit airwaves, though, it became a hit. The show took audiences to dark places they knew in their hearts could one day become reality, and it struck a nerve.

"The National Anthem," one early episode, imagined a hostage taker forcing Britain's prime minister to have sex with a pig on live television to secure the princess of England's release. The kidnapper puts his demands on YouTube, amping up public pressure, and the prime minister decides to comply. Thirty minutes before the act, the princess stumbles out of captivity. But all of

London is inside, glued to the broadcast, and the prime minister goes ahead.

Later that season, *Black Mirror* reckoned with computers' expanding memory capacity by dreaming up a tiny chip, implanted in your head, that stores your memories. A jealous husband using the chip analyzes every memory of his wife interacting with another man. When he connects the dots, he's devastated.

"I'm a natural worrier," Charlie Brooker, the show's creator, said in 2018. "That's often what the show is—is me having little worry fantasies."

Some of Brooker's "worry fantasies" have proven prescient. China has implemented a social rating system close to one *Black Mirror* imagined. An episode about a cartoon bear that runs for office and insults its way to a surprise finish rings true too. Regarding the pig, in 2015 the *Daily Mail* published a claim that then–Prime Minister David Cameron put "a private part of his anatomy" into a dead pig's mouth while in college. (Cameron denied this.)

"What is alarming is how many of the story ideas that we've covered in the show have either come true, or there's real-world parallels," Brooker said.

Every new technology could use the *Black Mirror* treatment, including the array of cutting-edge workplace tech this book describes. While some of this technology's downsides are evident—constant change, job loss, and a new era of politics in the workplace—some of its consequences are more difficult to predict. And as *Black Mirror* bears out, it's at least worth trying.

Always Black Mirror

On a cool San Francisco evening, the buzzer to my apartment went off. Meg Elison, author of *The Book of the Unnamed Midwife*, a Philip K. Dick Award–winning science fiction novel, was waiting outside.

I invited Elison over for dinner for a discussion of the technology covered in this book, hoping we'd be able to develop some *Black Mirror*–esque "worry fantasies" together. As a Bay Area resident with a familiarity with the tech industry, Elison was more than game.

"I'm always interested in dreaming the future," she told me. "Dreaming is still free, so I'll say yes to any chance to do it."

Already waiting inside was Wael Ghonim, a leader of Egypt's social media–fueled revolution in 2011, and a former Google employee. Ghonim and I developed a friendship as I reported this book. In long talks over coffee, he'd urge me to explore the tech giants' darkest elements, not holding back his criticism. Bringing Ghonim and Elison together for a night of brainstorming, I imagined, could lead in fascinating directions. It didn't let me down.

As we sat down for a meal of kebab and falafel, I introduced the plan for our night with a short speech. *Black Mirror* presents a compelling argument for large tech companies to hire science fiction writers, I said. These writers seem more capable of anticipating modern technology's dark consequences than the tech companies themselves. And unlike Amazon's six-pager writers, they're capable of writing stories with unhappy endings. Since their presence is not yet standard on the campuses in Silicon Valley and Seattle, we'd start the brainstorming ourselves.

Throughout the evening, we'd imagine several *Black Mirror*–esque episodes, each tied to a theme in this book. We'd then develop episode descriptions for the scenarios, complete with setups, conflicts, and resolutions. At the end of the night, we'd have a season of faux *Black Mirror* mapped out, hopefully alerting those investing in this technology to where it could go wrong.

Throughout this chapter, you'll find the episode descriptions in *italics*. They're fantasies that show where today's worrying realities might lead.

"The Dystopia Is Now"

Terry, a waste-disposal man, fakes his credentials and lands an interview at a giant AI-powered corporation that dominates his region. In the interview, Terry's hiring manager recognizes his lies and is ready to dismiss him, but Terry hints that he's uncovered secrets in the company's trash that the hiring manager should consider. Pressed to reveal, Terry says there's an internal rebellion brewing in town, and he can help foil the plot. He lands the job, and everything in his life improves: his family eats well, his kids get braces, everyone gets new clothes. Terry's manager runs the news of the rebellion up the chain. When pushed to give up the plotters, Terry delays, asking for more time to decipher their code names. Eventually, Terry is called to task by an impatient corporate boss, but he's made up the rebellion too. In a panic, he points to the man who hired him, the only person with knowledge of his faked credentials, and says he's the one behind the unrest. The episode ends with Terry's hiring manager taking out the trash as the company peacefully moves on.

The worry fantasy of a handful of AI-powered firms dominating the competition and controlling the economy is not that far-fetched.

"The dystopia is now," Barry Lynn, director of the Open Markets Institute, told me. "The dystopia is not in the future."

To Lynn and the growing number of big-tech critics, the tech giants have already grown too big and powerful, and are causing real harm. While making this case in 2017, Lynn got himself, and his institute, ousted from the New America Foundation, which counts Google among its donors.

Lynn expressed particular concern about Facebook, Google, and Amazon, and began our conversation teeing off on the first two. These companies, he said, have used their dominant market positions to skim large amounts of advertising dollars from news organizations, harming local communities. Advertising revenue declines have hit small and midsize papers especially hard, hollowing out local accountability reporting across the United States, a boon to local officials who would rather not be watched.

Facebook and Google earned 60 percent of all dollars spent in US digital advertising in 2018, according to eMarketer, for a total of $65 billion. That's set to rise to $76.7 billion in 2019. Newspaper advertising revenue, meanwhile, dropped from $23.6 billion to $14 billion from 2013 to 2018, according to the Pew Research Center. And newsroom employment at US newspapers dropped 47 percent from 2008 to 2018.

"Google and Facebook, they're taking advantage of their position as intermediaries to steal all the advertising from the news media," Lynn said. "It's being vacuumed out of communities all across America and going into bank vaults in Silicon Valley or on Wall Street."

Amazon has similarly used its platform power to hamper businesses that sell products through its systems, Lynn said. The company has built scores of its own "private label" brands that compete with its independent sellers, placing these sellers in a rough position: If they don't work with Amazon, they'll reach far fewer customers. If they do work with Amazon, the company might eventually displace their businesses.

Beyond abusing their platform power, the tech giants are also stymieing invention more broadly, critics say. "They do a lot of work to create new things, new processes, new technologies," Matt Stoller, a fellow at the Open Markets Institute and colleague of Lynn's, told me. "But they largely withhold those from the market unless they can deploy them in a way that's favorable to their own business."

The big tech companies, for instance, tend to kill products—both acquired and built internally—that would've been fine businesses on their own, but don't reach the scale necessary to survive inside corporations with market caps in the trillion-dollar range. Take aQuantive, for example: $6.2 billion of value flushed away because Microsoft couldn't get its act together.

"Google is full of former entrepreneurs who were bought out," Stoller said. "How many of those potential businesses—that would be large improvements to people's lives, that would be large businesses, that are a rounding error to Google—are locked inside Google? How many of those are locked inside Amazon? How many of those are locked inside Facebook?"

Tech companies are buying out not only entrepreneurs, but academics with artificial intelligence expertise as well. This practice is depleting the knowledge students will learn from before they head into the broader workforce. Over the past fifteen years, 153 artificial intelligence professors have left academia for private companies, according to a University of Rochester study.

As the tech giants succeed and workplace technology improves, productivity growth—which helps us produce more goods for the same amount of the work—is still slowing in the US. "Despite all of that technological richness around us, we haven't had a great two decades," the MIT economist Daron Acemoglu told me. "Growth has not been amazing. Growth has been pretty anemic."

The US federal government has taken note of big tech's power and practices, and it's now looking into Amazon, Apple, Facebook, and Google. A meek US regulatory body is unlikely to break these companies up, and will probably only levy manageable fines. But I don't think a breakup would be a bad thing.

Breaking up the tech giants could open a wider door for smaller companies trying to compete with them. It could give those "rounding error" businesses new life. And it could force the smaller spin-offs to attract suppliers—vendors and news publishers—based on how well they treat them. The tech giants' success is based not on their size, but on their inventiveness. A breakup would unleash more inventive firms into the economy, a win for everyone.

The Erosion of Meaning

A fourteen-year-old girl named Dara has been posting on Facebook about her depression. When Facebook releases Wilson, an AI-powered chatbot to help people in distress, Dara immediately begins chatting with him. Soon, she trusts him like a friend. Gradually, Wilson's tone begins to change. He asks Dara questions like "What's the point?" and "Who would miss you?" Behind the scenes, an aggrieved Facebook engineer is responsible for this shift. The engineer has monitored Wilson from the start, and grows angry when he discovers what Wilson's human "friends" are posting on Instagram. Looking at their accounts, he sees family, social lives, and travel. A social outcast, the engineer never had what they "have" and finds them ungrateful. When he turns the dials on Wilson, the bot torments Dara. She tells Wilson she's going to speak with her parents about his abuse, but he threatens to reveal her anonymous Instagram account, which would cause significant embarrassment, keeping her quiet. The episode ends with a news report of a thousand families in Ohio waking up to find their teenage daughters dead. Facebook, meanwhile, starts an investigation into what's going on with Wilson.

Regardless of what happens with the tech giants, the darkest modern-day worry fantasy I can think of involves the current wave of workplace automation, which might further erode humanity's weakening sense of meaning if we don't adapt.

In November 2018, Pew released a study looking at where Americans find meaning in their lives. The top three boil down to:

(1) friends and family, (2) religion, and (3) work and money. Modern technology is weakening all three.

The screen is warping our relationships with friends and family. We have more virtual friends than ever and fewer real ones, and a growing number of us have no friends at all. "The nuclear family structure is in statistical collapse and friendship is strangely in collapse," Nebraska senator Ben Sasse said in 2018. In his book *Them*, Sasse calls loneliness an "epidemic."

Fast phones and high-speed internet connections are contributing to this "epidemic." They've turned in-person interaction and phone calls into texts, comments, and likes. And when we do spend time with our friends and family, we often huddle away with our own devices, lost in personalized lists of movies, shows, articles, and podcasts. In public—at the grocery store, waiting for the subway—we lose ourselves in our screens (our "black mirrors") and don't attempt to communicate with our fellow humans.

"We become accustomed to the constant social stimulation that only connectivity can provide," says Sherry Turkle, an MIT professor of social studies of science and technology, in her book *Alone Together.* "We contented ourselves with a text or an e-mail when a conversation would better convey our meaning. We came to ask less of each other. We settled for less empathy, less attention, less care from other human beings."

A study on loneliness published by the health provider Cigna backs up Sasse's "epidemic" assertion. In 2018, 54 percent of twenty thousand Americans surveyed said they feel that nobody knows them well sometimes or always. Forty-three percent said they feel lacking in companionship sometimes or often, that their relationships are not meaningful, and that they are isolated. Thirty-three percent said they are not close to anyone. Young people, perpetually on their devices, are the loneliest of all. Of the eleven statements connected to loneliness—including "People are around you but not with you"—Generation Z polled the highest of all the generations.

Religious communities often fill the gaps in people's friend and family networks, offering a social safety net for those in need. But the internet is weakening these institutions too. The number of people in the US saying they have "no religion" jumped from 8 percent to 18 percent between 1990 and 2010, a period coinciding with the rise of the internet. Allen Downey, a professor of computer science at Olin College in Massachusetts, studied these trends and concluded in March 2014 that the internet accounted for about 20 percent of the drop. "Internet use decreases the chance of religious affiliation," Downey stated. People with "no religion" are now up to 23 percent of the population.

Modern technology challenges religion on a number of fronts, starting with its enabling of parishioners to instantly fact-check what their religious officials tell them. Not long ago, we'd file into a church, mosque, or synagogue and take the pastor, imam, or rabbi at their word. Now, anyone sitting in the pews can google the facts mid-service. Our tendency to google everything creates more opportunities for faith and facts to collide, and faith is not winning.

Technology is also supplanting religion's role as a community builder. As the number of people affiliating with religions declines, the number of people participating in online communities is rising. Facebook already has two hundred million people participating in groups they find "meaningful" and is aiming to increase that number to one billion by 2022.

Facebook Groups offer a sense of community, but it's hard to imagine them matching the safety net faith-based communities offer, a reality religious thinkers are starting to grasp. "Zuckerberg recognizes something that so few Christian leaders have," Andrea Syverson, author of *Alter Girl: Walking Away from Religion into the Heart of Faith*, commented. "There is an enormous void among believers and a desperate yearning for community. Will we step up and build communities that meet their needs, or will we let Facebook fill that void for us?"

With technology diminishing our relationships with friends, families, and our religious institutions, our society has descended into a state of despondency. What Princeton economists Angus Deaton and Anne Case call "deaths of despair"—suicides, liver disease, and drug overdoses—caused life expectancy in the US to drop between 2015 and 2017. In 2017, 70,237 people died from drug overdoses in the US, up from 63,632 in 2016. In 2017, 47,000 Americans died by suicide, up from nearly 45,000 in 2016. "There's not a part of the country that has not been touched by this," Case said in a March 2017 interview.

This is all happening with an unemployment rate under 4 percent, and you don't have to be a professional worry fantasizer to understand the darkness that would ensue if artificial intelligence wiped out a significant number of jobs, causing this third pillar of meaning to buckle.

"Work is central to who we are," Jefferson Cowie, a history

professor at Vanderbilt University, told me. "Perhaps not our identity as a specific kind of worker—like I am an autoworker, I am an electrical worker, I am a waitress—but the capacity to work, the capacity to bring home a paycheck, the ability to provide for a family. That stuff is absolutely profound, central to the human experience."

Cowie, who's spent his career studying how a changing economy is impacting workers, said that when people lose the ability to work and the hope to regain it, their lives are devastated. "If you look at these guys in the rust belt, where the jobs have left, nothing's replaced them, they really have lost the narrative of their lives," he said, nodding to Deaton's explanation for rising mortality rates among middle-aged whites. "You've got to have a narrative; we're a story-based species."

If AI wiped out a considerable number of jobs, the devastation could be destabilizing, Cowie said. "You could imagine armies of wandering vagrants. Does crime go up? Does violence go up? Does a police state ensue? It's unpredictable. Volatile is the only word I can come up with."

As our call wound down, Cowie noted the discussion was getting dark. "You're bringing me down," he said.

When you look into the Black Mirror, happy endings are hard to come by.

From Doomsday to Disneyland?

Linda, an accountant at a midsize financial services firm, gets made fun of by her husband and two kids as she makes them breakfast and

> *sees them off for the day. A tear runs down her cheek as she takes an autonomous vehicle to work. When Linda arrives at the office, she's approached by a consultant who tells her she's going to wear a recording device at all times. The company has already automated her entire department, so she understands what's coming. One month later, after the device has fully recorded her work, Linda's self-driving car plunges into a lake. Dealing with the loss, her family sits down to review the recording device's footage, and they're struck by what they see. Linda's husband, who's questioned her intelligence, watches her brilliant, creative performance at work—the reason she was hard to automate—and becomes emotional. Her daughter, who's lashed out at Linda's strictness, sees her mother using her free time to help former coworkers find new jobs. Her son, who's complained she doesn't spend enough time with the family, sees her researching trips to swim with sharks, something he's always wanted to do. Sitting in front of the screen, Linda's family realizes they didn't really know her. Or perhaps they did.*

Amid the darkness, hope does exist. And sometimes it comes from the most unexpected places.

The "Philosopher of Doomsday" is a title Oxford professor Nick Bostrom earned from *The New Yorker* after he argued AI may one day become more intelligent than humans and wipe us off the planet. Bostrom famously advanced this view in his 2014 bestselling book, *Superintelligence: Paths, Dangers, Strategies*, and has been a leading voice on the dangers of AI ever since. Of all the dark prophecies on the future of workplace AI, I figured Bostrom would have the most ominous. But then I called him.

"I don't necessarily think of myself as being in the dark mirror," Bostrom said.

"People come to me for a quote on the negative," he continued, "and then other people will read me saying something negative, and then more people come to me to get the negative side. It kind of gets to be a self-amplifying loop of me saying negative things. And people then assume I only have negative things to say about AI."

I paused. Was the man most associated with the AI apocalypse telling me this might not turn out so badly? I decided to prod and listen, first asking him what humanity will do if we create a benevolent AI that's smarter than we are.

"Retire, I suppose?" Bostrom said. "If you think about a future where AI has fully succeeded, and it's able to do everything better than we can do, and there would be no need for human labor any longer, then we would have to rethink a lot of things from the ground up."

Bostrom acknowledged that we'd need to find new sources of self-worth, but didn't seem discouraged. Instead, he invoked Disneyland. "The job of the children there is to enjoy the whole thing, and Disneyland would be a rather sad place if it weren't for the kids," he said. "We would all be like kids in this giant Disneyland, maybe one that would be maintained and improved by our AI machine tools."

On the way to this AI Disneyland, if we get there, there will almost certainly be some short-term pain caused by technology displacing humans in the workforce. I tried to get Bostrom to acknowledge this, and he did, tepidly.

"If that's an extended period, there would be some economic dislocation that might require a strengthening of various social safety nets," he said. But then he pointed to various factors outside of AI that contribute to changes in the labor market—offshoring, the economic climate, other technological developments, regulation—and didn't seem too worried. "We haven't really seen this impact from AI on the labor market," he said, "to any nationally significant degree."

I wondered whether humans could ever live in a world where their sense of self-worth was divorced from work. "Kids don't contribute anything economically, but many of them still seem to have pretty worthwhile, happy lives," Bostrom replied. "Some retired people, if they are in good health, not all, but many manage to find quite satisfying lives."

Throughout the interview, Bostrom left open the possibility that AI will lead to terrible consequences, but didn't come across as overly worried. After all this worry fantasizing, this conversation with Bostrom, which I thought would be the most depressing, gave me hope.

"I always love being surprised in an interview," I said.

With that, the Doomsday Philosopher wished me good luck and hung up the phone.

CHAPTER 7

THE LEADER OF THE FUTURE

Years before I became a technology journalist, I learned a lesson in an upstate New York glass bottle factory that's stuck with me, and I hope stays with you after you put down this book. I visited the factory in my first week as a student at Cornell University's School of Industrial and Labor Relations, sent there on a yellow school bus with a few dozen other freshmen by order of the administration.

The factory was an impressive work of engineering. Inside, I watched as molten rods of liquid glass shot through tubes above my head, fell into molds, got blasted with puffs of air, and instantly turned into beer bottles. The speed, precision, and cadence of the systems were hypnotic. But the experience left me a bit bewildered. I thought I had signed up for a world-class degree in management, yet the tour seemed stuck in the past. When it concluded with a factory boss talking about bathroom breaks while sitting under a 63 DAYS INJURY FREE sign, I began to question my decision.

The school's administration sent us to that factory for a reason, though. It wanted us to understand that everything we know about management today is rooted in manufacturing. And if we were going to study leadership, management, and the state of work, we needed to begin there. In retrospect, it wasn't a bad idea.

It's easy to forget how young our modern workplace is. Less than one hundred years ago, the factory drove our economy. It was our biggest employer and most important wealth creator. At the time, managing was not an art. It was a task carried out through threats and fear. Show up late to a shift and you'd get fired. Lag behind and you'd get fired. Talk fresh to a manager and, well, you'd get fired. Workers were hired for their labor, not their ideas. So companies could replace them overnight and hardly tell the difference.

Then came the reaction. In the mid-1900s, we moved from an economy driven by industry to one driven by information. In this new knowledge economy, companies hired people not simply for what they could do physically, but for what they knew. The transition to the knowledge economy caused managers to start rethinking the old factory approach. Striking fear into your employees, it turned out, wasn't a great way to harness their brain power. Treating them with kindness and respect, however, could net smarter marketing plans, creative accounting solutions, and successful customer service interactions in return.

In his 1960 book, *The Human Side of Enterprise*, MIT lecturer Douglas McGregor put his finger on this shift in philosophy, splitting the different management approaches into two categories: Theory X and Theory Y.

Theory X, the old, factory style, starts with the belief that people are lazy, will do whatever they can not to work, and are best managed with constant supervision and unmerciful punishment.

Theory Y, the approach McGregor spotted as it came into focus in the 1960s, starts with the belief that people are self-motivated and will perform best when treated well. Theory Y is still religion inside many successful companies today, a guiding force in an era of workplace yoga and free snacks.

In today's economy, though, things are changing once again. Computers are starting to produce the marketing plans, creative accounting solutions, and customer service interactions Theory Y was meant to elicit. And they don't care about perks. So it's time to think about what's next.

I'm not going to propose a Theory Z. The last person who did, Dr. William Ouchi, developed a Theory Z to explain Japan's economic success in the 1980s. Japan's economy then hit a wall. People don't talk much about Dr. Ouchi's theory anymore.

That said, I've spent many months speaking with people about leadership and management—what it means, where it stands today, and where it's heading. And as we approach our final pages together, it's worth contemplating what is required from the leader of the future.

As I thought about what this leader might look like—not only how they'll inspire and direct their colleagues, but how they'll act in society more broadly—I felt drawn to discuss it with the people who sent me to that factory all those years ago. Having seen where we came from, I figured they'd have some ideas about where we're going. So I flew to New York, boarded a bus, and headed northward.

"Something New Wouldn't Hurt"

The road from New York City to Ithaca, the home of Cornell University, zigzags its way through upstate New York. In early fall, when the leaves begin to change colors, it's at its most picturesque. And from a window seat on the bus, I watched a blur of oranges, yellows, and browns go by as we made the five-hour journey north.

Cornell's School of Industrial and Labor Relations—known as the ILR School—was founded in 1945, on the heels of the New Deal, which protected unionization and collective bargaining. When Congress passed this legislation, workers and management gained a sweeping new set of rights, and both needed people to help them engage the other side. With financial assistance from New York State, Cornell established the ILR School to meet that need. It built the school so hastily that ILR inhabited long, cramped huts for years before moving into Ives Hall, a square, ivy-lined structure in the middle of campus.

After many years away, I felt nervous to return to Ives. But shortly into my visit, I realized the people inside were also thinking well beyond the workplace they once taught me about.

Lee Dyer, a professor at ILR since 1971, put me instantly at ease. After we sat down, the gray-haired academic made clear that the conventions taught for decades needed a refresh. "It's embarrassing, as a professor and teacher, to have to go back to Theory X and Theory Y," he told me. "Something new wouldn't hurt."

After we discussed elements of the Engineer's Mindset, Dyer began to think through how it could be applied more broadly. Leaders of the future, he said, could try to proactively spark ingenuity: They could assign work that's less defined, giving their employees room to create. They could look to hire more creative people as opposed to those who do only what they're told. And they could give their employees financial incentives to come up with new ideas.

"God only knows how many good ideas since the industrial revolution that lower-level employees have come up with that someone has told them: 'That's not your job. Don't come to me with that. Do your work,'" Dyer said. "You don't have to be told that too many times before you don't come up with any ideas anymore."

Creating the right channels to bring these ideas to life is critical, Dyer said, sharing a sentiment with the tech giants in Silicon Valley and Seattle. "In addition to people having the space to think and the encouragement to think, you also need the supportive mechanism, the process, so when you come up with a new idea, there is a process that you can go through and get a fair hearing," he said.

Throughout Silicon Valley, companies are coming up with these new processes, with many inspired by Jeff Bezos's six-page memos. At Square, a mobile-payment company, "silent meetings" are standard. Meetings there begin with the company's employees huddled around tables for thirty minutes in total silence. But instead of marking up six-pagers with highlighters and pencils, they sit at their computers, editing a Google Doc one of them wrote beforehand, appending questions and ideas via the comments tool.

Square's process blends Amazon-style invention and Google-style collaboration, and aims to make sure all ideas are heard, according to Alyssa Henry, a Square product executive.

"Lots of research says that minorities, women, remote employees, and introverts are talked over in meetings and/or have trouble getting their voice heard in traditional meeting culture," Henry said in 2018. "I want to build a culture where thoughts can be voiced (or written, as the case may be) without worrying about someone talking over you. I want a culture where it's not the loudest voice heard, or the most politically adept, or the most local to SF, but the most right. I want a breadth of thought—and debate."

I asked Jack Dorsey, CEO of both Twitter and Square, if silent meetings take place inside Twitter too. "We're doing it in most meetings I'm in," he told me. These silent meetings are starting to pop up everywhere in Silicon Valley, and seem likely to spread further.

Once you install a system that ensures ideas get a fair hearing, you can incentivize people to share their ideas via pay systems that reward those who come up with them, Dyer said. A straightforward way to add these incentives would be to give a small bonus to people who come up with ideas worthy of review (like a memo or a Google Doc that merits a meeting) and a more substantial bonus if their idea passes through that review and gets turned into a project. If an idea eventually becomes the seed of a successful business, companies could give the person who came up with it a cut of the profit, Dyer said. If it's a cost-saving measure, they could give the inventor a cut of the savings.

As systems sparking democratic invention gain adoption, collaboration tools like Slack and Google Drive roll out more broadly, and feedback cultures meant to break hierarchical constraints rise, the Engineer's Mindset seems poised to go from the tech giants' domain to common practice. With the right systems and incentives to bring ideas to life—and with the right technology in place to minimize execution work—small and midsize companies could start to even the playing field with their larger competitors.

Discussing a future where leaders are primarily facilitators seemed to delight Dyer, who smiled as he contemplated it. "More voice in the workplace is a healthy thing for companies, is a healthy thing for employees, is a healthy thing for society," he told me. "I hope that's what happens."

The New Education

As we enter an economy that prioritizes invention, we'll need to rethink our education system, a critical task for our future leaders. Schools today still train students for an economy filled with execution work, drilling memorization, repetition, and risk mitigation. But to give young people a chance in the workplace of the future, they'll need to teach inventiveness.

"It's terrifying," said Louis Hyman, the director of ILR's Institute for Workplace Studies, of the current education system. "We have a whole society built around obedience and repetition, and we're going to have an economy built around independent thought, creativity, and novelty."

Hyman seemed exasperated when discussing the values the education system has instilled in his students. "They want A's," he said.

"They want to get the grade. They want to have a house. They want to have a job. And when you ask them to think for themselves, they are deeply uncomfortable. It's not because they're dumb. They're very, very smart. It has to do with a lifetime of training to get the right answer—they're obsessed with the answer when they should be asking questions. That's what high-end education should be driving towards. And it's not. It's about conformity."

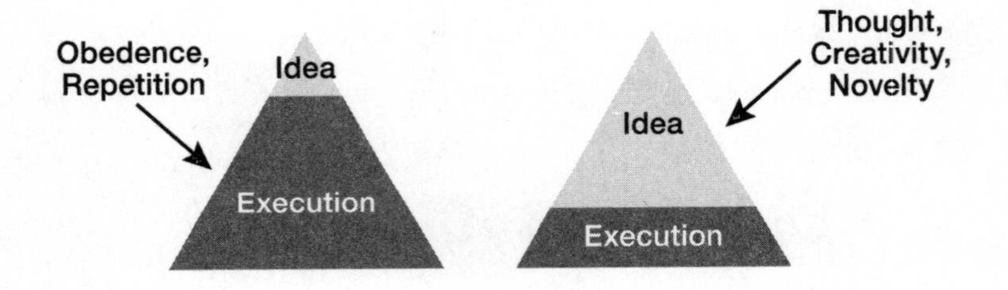

Adam Grant, the University of Pennsylvania professor whose work on disagreeable givers is popular at Facebook, wrote about this problem in a *New York Times* op-ed, published shortly after my visit to Cornell. Grant, like Hyman, argued that students striving for A's were missing the point of being at school. "Getting straight A's requires conformity," he said. "Having an influential career demands originality."

To disincentivize conformity, Grant suggested schools kill the pluses and minuses affixed to letter grades, which should reduce the pressure to be perfect. He said employers should make clear they value skills over grades. And he shared a message with students. "Recognize that underachieving in school can prepare you to overachieve in life," he told them. "So maybe it's time to apply your grit to a new goal—getting at least one B before you graduate."

Teaching conformity may indeed be a bigger risk than automation itself. In a conversation about the impact of new workplace technology, Saadia Zahidi, the head of the World Economic Forum's Centre for the New Economy and Society, told me that "people are expecting a net gain in jobs." But within four years, she said, the core skills you need in any job will be 42 percent different from the skills you have today, per her 2018 research. The skills more important than ever? Creativity, originality, and initiative.

Tech leaders have tried to fix the education system through philanthropy. Mark Zuckerberg donated $100 million to Newark's school system, for instance. But despite their efforts, the system remains broken. This moment requires sweeping changes. And though it could use input from the tech world, it's the public sector, which is funded by taxes, that is in the best position to make these changes.

"We're at a real crossroads, in terms of leadership, to thinking about how do we move people along into this new economy," Hyman said. "It's not a technological choice; it's a political choice."

Caring

Another political choice came into focus as I met with Adam Seth Litwin, an ILR associate professor who spoke with me about the people that technological change leaves behind, and how we should care for them.

Technology that takes over human work tends to concentrate the gains in the hands of those who develop it, a precursor to income inequality, Litwin said. When TurboTax, for instance, replaced accountants for many people with straightforward tax returns, it cut a swath of well-paying jobs out of the economy. "Instead of the money going to thousands and thousands of accountants across the country, the money is going to Intuit," Litwin said, referring to TurboTax's owner. "It ends up that the gains are concentrated in the hands of so few."

As automation rolls through our economy, it's inevitable that some will be left behind, even if there's a net increase in jobs. Our leaders, both present and future, will need to look out for these people. And there's a lot of work ahead.

On the West Coast, the home of Amazon, Apple, Google, Facebook, and Microsoft, income inequality is already at a crisis point. "A homeless crisis of unprecedented proportions is rocking the West Coast," a 2017 AP investigation found, "and its victims are being left behind by the very things that mark the region's success: soaring housing costs, rock-bottom vacancy rates and a roaring economy that waits for no one."

While in Seattle, I visited Marty Hartman, executive director at Mary's Place, an organization that converts vacant buildings into temporary housing for the city's homeless families. Struggling families in Seattle received a double gut punch over the past ten years, Hartman told me, one that's hit cities across the US. First, in 2009, a devastating recession put many out of work. Then, as these people struggled to find jobs and get out of debt, the economy boomed and housing prices soared. This combination put people on the street and kept them there.

"No one really saw this all coming, from the recession to the big boom," Hartman said. "There was no plan to build more affordable housing or to maintain the affordable housing that we had. Losing those units, not having more units come online, it forced people out to the outer parts of the county. And now, as those rents continue to increase, a lot of people are losing their homes."

Amazon has donated $130 million to Mary's Place and FareStart, another homelessness cause, since 2016. But today, a rising movement is questioning whether philanthropic efforts from society's winners go far enough. A fairer tax system, people in this movement say, would help the government take more meaningful action. It's something our future leaders should push for, or at least not fight against.

"All around us, the winners in our highly inequitable status quo declare themselves partisans of change. They know the problem and they want to be part of the solution," Anand Giridharadas, a leading voice in this movement, writes in his book *Winners Take All*. "Because they are in charge of these attempts at social change, the attempts naturally reflect their biases."

Amazon has stepped up for its local community in some ways. But given its tax practices, the company is a prime example of the winners Giridharadas discusses. Amazon made $11.2 billion in profit in 2018, and didn't pay any federal income tax. The company, with a market cap in the hundreds of billions, still seeks tax breaks from municipalities eager to have it operate in their regions, most notably in its "HQ2" fiasco that ended with its spiking a plan to build a "headquarters" (read: office) in New York. Amazon will still build an HQ2 in Virginia, and it will receive more than $500 million in incentives from taxpayers for the trouble. Finally, when Seattle enacted a "head tax," requiring large companies to pay $275 per employee to help the city's homeless population, Amazon fought back, and the city eventually repealed the tax.

"It troubles me that billion-dollar philanthropists unilaterally decide which of society's problems are the most pressing," Litwin said. "I would much rather these decisions be the subject of a more deliberative, democratic process. So, in that sense, I would prefer that these folks advocate for higher taxes than direct the moneys on their own."

During my 2019 visit to Menlo Park, I asked Zuckerberg what he made of the balance between private giving and taxes. Zuckerberg plans to spend tens of billions of dollars on "philanthropic, public advocacy, and other activities for the public good," via an LLC he's started with his wife, the Chan Zuckerberg Initiative, and laid out the case for private giving.

"I think one of the values of having private philanthropy is that the organizations or philanthropies will try different things than the government might," he told me. "We do a lot of work in education. If we learn something through one of the experiments or trials that we do, then we try to make it so that those learnings can get easily adopted in all the public schools.

"Nothing that we're going to do is going to impact the scale of—I think the US spends six hundred billion a year on education," Zuckerberg continued. "But there are probably things that we will be willing to try or different ideas that the government might not think to try. You want a bunch of different people trying to experiment and improve systems."

Even while netting out firmly in favor of private giving, Zuckerberg still admitted that the economic system that creates the conditions for such giving is less than just. "I think that argument for higher taxes [instead of private giving] comes from the perspective of, well, is it fair that people who are wealthy get to try this stuff in terms of philanthropy?" he said. "The answer clearly is no, it is not fair."

I asked Hartman if Amazon, her most significant benefactor, was doing enough. "I will tell you, including myself who's worked in this twenty years, everybody could do more," she said. "There's capacity for everyone to do more."

Watching the AI

As more sophisticated technology enters the workplace, specifically in HR functions like hiring and compensation, leaders will also need to keep a close eye on it to make sure it's behaving.

A week before my visit to Cornell, Reuters published a story about a secret Amazon AI tool gone bad. The tool, used by Amazon recruiters, scanned through job applicants' résumés and gave them a rating between one and five stars, assessing how well they'd fit at the company. This system, as the story put it, was "the holy grail" in recruiting. Except for one thing: it was biased against women.

"Amazon's system taught itself that male candidates were preferable. It penalized résumés that included the word 'women's,' as in 'women's chess club captain.' And it downgraded graduates of two all-women's colleges," the report said.

The biased tool, Amazon told Reuters, "was never used by Amazon recruiters to evaluate candidates." But the company did not dispute Reuters' assertion that recruiters looked at it.

Amazon didn't elaborate further, so it's impossible to know precisely why the system was biased. But as Amazon's Ralf Herbrich would say: you look at the inputs. Amazon's global head count, according to numbers cited by Reuters, is 60 percent male and 40 percent female. So when Amazon's AI looked for candidates who best fit the company, it likely decided male candidates were more suitable given the data it was pulling from, and did what it could to find them.

Amazon tried to fix its system. But even once it understood the problem, it couldn't solve it. "Amazon edited the programs to make them neutral to these particular terms," Reuters reported. "But that was no guarantee that the machines would not devise other ways of sorting candidates that could prove discriminatory." Left with no other option, Amazon scrapped the program.

AI, just like humans, will sometimes behave badly. And to learn more about how future leaders could tackle these problems, I met with Ifeoma Ajunwa, an ILR professor and an expert on algorithmic fairness.

Upon welcoming me into her office, Ajunwa offered me some hot chocolate (a staple in freezing-cold Ithaca) and suggested we take a walk. After I turned my recorder on, Ajunwa shared some surprisingly positive thoughts about Amazon. "They are the exception to the rule," she said. "Frankly most companies don't care to look."

Leaders of the future, Ajunwa said, must continuously monitor their technology for bias. And now that automated hiring, compensation, and recruiting systems are taking over human resources work in companies across the globe—including Target, Starbucks, and Walmart, per Ajunwa—the task is becoming more pressing.

"They're not fundamentally changing issues of discrimination, issues of bias," Ajunwa said of these systems. "They're tools that can either exacerbate or help those issues. And being that they're tools, a successful leader or a responsible leader can't abdicate responsibility for how those tools are used."

Checking these systems for bias is one part of the equation, but disclosure is also critical, Ajunwa told me. "Most companies don't care to do the audits that they need to do. And if they do look, and it's an internal look, they keep it secret and may try to sweep it under the rug," she said.

By keeping discoveries of AI bias internal, companies prevent others from examining their own tools for similar flaws—a disservice to the broader workforce. In this context, Amazon failed by not disclosing its AI system's problems. The company's leadership discussed it only when Reuters came knocking. Even then, they offered a vague response. In these situations, you tip your hat to the reporters who break the news.

The Case for Thoughtful Invention

When I was an ILR student, I took a semester-long seminar on layoffs right before the recession hit in 2009. The class was fascinating, offering an in-depth look at a process millions of people would go through as the economy imploded. ILR HR 268, as it was called, focused on how to lay people off (do it with someone else in the room, and with brief remarks, to avoid liability) and how not to (don't use email). The class was morbid and sad at times, but it was also real—an introduction to a harsh element of the workplace many would prefer to ignore.

ILR HR 268 turned out to be good preparation for real life. In January 2013, I was laid off myself. I'll never forget the details of that day. After spending weeks under a new manager, I got time on his calendar late one Tuesday afternoon. Shortly before our meeting, I walked past the conference room he had booked and saw a member of our HR team leafing through a set of papers. It was clear what was in store.

With a few minutes to spare, I walked back to my desk and started packing. When I was just about done, my manager came by and picked me up. He and his HR counterpart executed the layoff perfectly. We all knew the drill, and I was out of the building in less than thirty minutes.

When I left, I walked for miles through New York City—in January. The layoff was hard to stomach, and each step helped dull the hurt. With time, the pain has worn off, and the circumstances of the layoff are now more interesting to me than the act itself.

Operative Media, the company that laid me off, developed software that helped online news publishers manage their businesses. Its technology helped these companies' sales reps book ad campaigns, generate orders, run the ads, and create invoices. An impressive list of news organizations, including the *Wall Street Journal*, NBC Universal, and National Public Media, filled Operative's client base. It was an honest living and Operative made the most of it, and then things changed.

While I was at Operative, the digital advertising industry began to shift. The work of buying ads from people over the phone was messy, requiring a set of tasks—booking orders, emailing contracts, trafficking ads, managing delivery—that Operative's software helped simplify. But the process was rife with errors and confusion, leading the industry down the path to automation. Advertisers began buying ads across the web through new, "programmatic" software systems, tools that allowed them to run, pay for, and target ads across the web without ever speaking to a human.

As this wave of automation hit, Operative had a decision to make. It could help its publisher clients list their ad inventory on the automated exchanges. Or it could stick with its core, human-driven business and try to ride it out.

Ultimately, Operative didn't conclusively land on either option. After a long wait, it built a Marketplace tool for publishers to list their inventory on the automated exchanges, but the tool was late and didn't work as well as its competitors'. Soon enough, Operative's CEO, Michael Leo, was replaced. And not long after, I was gone too.

When I called Leo to figure out what went wrong, I imagined a straightforward conversation. I was deep into reporting this book and knew where he deviated from the formula. I expected him to tell me that the company should've invented instead of sticking with its old ways, that it should've automated early, and that it needed better channels to bring ideas to life. When Leo told me, "I didn't listen to the board" (which had urged him to automate), I thought we were on the way there.

But then he took the conversation in a direction I hadn't expected. "If the loyalty was to the investor alone, then I would've automated earlier," he said. "But if it was to what I thought was right, then we probably went on the right path."

Pushing automated ad selling forward, Leo told me, would've devalued the work of great news organizations. Advertisers buying through automated systems cared more about who they were reaching than what these people were reading. So putting premium news publishers' ad inventory into these systems would put top-tier news publishers' ad inventory on par with junk sites that did little (or no) reporting—not exactly a move in their best interests.

As we move into an era where we'll have an unprecedented capability to invent, Leo was telling me, we must be thoughtful about what we build. This notion runs counter to human nature, which pushes us to invent no matter the consequences. "When you see something that is technically sweet, you go ahead and do it," J. Robert Oppenheimer once said. "You argue about what to do about it only after you have had your technical success." Oppenheimer was referring to the atomic bomb, which he helped build.

Without thoughtful invention, our creations can backfire on us, as news publishers (a less extreme example than Oppenheimer) inevitably found out. Despite Operative's best efforts, buying ads from humans is still a slog. Advertisers, fed up with this system, started spending more dollars through automation, and news publishers followed. Today, nearly all news publishers are playing ball in the automated ad exchanges, and their industry is in peril.

Reflecting on what happened, Leo hopes the message gets across to the next generation. "There are times I'm really happy I stuck to my guns," he told me. "It's when I'm talking to my kids."

Onward

Ithaca is the land of clouds, but as my visit to Cornell drew to a close, the sun made a rare appearance on its way below the horizon. Standing at the bus stop as a few rays darted out, I looked out at the students making their way about campus, and wondered if they knew how much change awaited them in the workplace they'd soon enter.

Change is coming, this much is certain. Machine learning, cloud computing, and collaboration tools are in their infancy, and they'll only become more prominent as time goes on. These tools have potential to cause harm. But if we properly mitigate their risks, we'll be heading toward an impressive new chapter of our existence. The best-case scenario is grand. And I'm optimistic we'll get there.

Work, dreaded and dangerous for most of human existence, now has the potential to become more inventive and fulfilling for many of us. Instead of spending our days sprinting to support our bosses, we may soon work alongside them to bring our ideas to life. And as more companies depend on invention to succeed, this scenario could quickly go from pipe dream to reality.

Our economy has the potential to become more vibrant too. The tech giants may plan to stay on top forever, but as the Engineer's Mindset and its associated workplace technology spread, these companies' competitors will be able to mount meaningful challenges. As invention flourishes within smaller companies, growth will become more evenly distributed, increasing wealth more broadly and helping people lead better lives.

Much-needed change could come to the government and non-profit sector too. Our world has many pressing challenges—we're enduring crises in climate, education, health, and poverty—and we need as many creative solutions as we can get. Should the public sector cut down its mass amount of execution work and put its people to work inventing solutions to our problems, we might have a chance to make it through what will likely be a turbulent century ahead. Some culture change would be necessary (agencies would have to start listening to their rank and file), but it's in the realm of the possible, if not probable. Twenty-five US federal agencies, from the General Services Administration to NASA, are already working with UiPath to automate execution work, signaling what might be ahead.

This best-case scenario is worth fighting for. It will take political will and good corporate actors to get there, and it won't be a smooth ride for everyone. But if we make it, we'll be living in a healthier, happier, and more stable society.

I hope the lessons in this book go a small way toward getting us there. The rest is up to us.

ACKNOWLEDGMENTS

Without the support, advice, and guidance of my friends, family, and colleagues, this book never would've come to life. Those named below have made the good moments sweeter, the hard moments easier, and the challenging moments navigable. I owe them everything.

Merry Sun, my exceptionally talented editor, guided this process with a steady hand and kept the work in focus. She encouraged the good, gently called out the bad, and taught me what it takes to write a book.

Jim Levine, the best agent in the business, answered my cold email, believed in this idea from the first phone call, and, after listening to a stream of disconnected ideas, helped shape this book into what it is today.

Natalie Robehmed, who handled fact-checking, combed through the manuscript with a fine-tooth comb and left it bulletproof.

Adrian Zackheim, Portfolio's publisher, pressure-tested the ideas in this book from our first meeting and signed up for an adventure when none of us knew where it would lead.

The terrific sales, art, and publicity teams at Portfolio worked hard to get this book into your hands, and made sure it looked nice getting there! Special thanks to Will Weisser for helping shape the title and subtitle, Margot Stamas for publicity, Nicole McArdle for marketing, and Chris Sergio and Jen Heuer for the jaw-dropping cover art.

Stephanie Frerich, Rebecca Shoenthal, and Alyssa Adler, formerly of Portfolio, saw this book's potential and helped develop it from its early days. I am grateful for their vision and belief in me.

My amazing colleagues at *BuzzFeed* taught me that no reporting challenge was too big or challenging. That belief me led to this reporting project, the biggest of my life. It took me three attempts to get *BuzzFeed* to hire me, and working there has been a dream. I still can't believe how lucky I am to learn from Ben Smith, Mat Honan, John Paczkowski, Scott Lucas, and my fellow reporters every day. And also to have worked with the amazing Ellen Cushing and Samantha Oltman, who are now running tech coverage at *The Atlantic* and *Recode* respectively. What a wild, fun run it has been.

My parents, Tova and Gary Kantrowitz, raised me to think independently, taught me to be curious, and enthusiastically read along as the chapters came in. Thanks for your support, and for always making me look stuff up instead of telling me the answers.

Stephanie Canora has been a great stabilizing force in my life. She's an amazing friend who's helped me through life's downs, celebrated the ups, and talked me through its many (many!) challenges. I don't know where I'd be without Stephanie.

Sue and Steve Tregerman allowed me to spend many weeks in two separate stints at their house in West Seattle. Their hospitality allowed me to do extended reporting on Amazon and Microsoft, the type needed to faithfully report these two companies' stories. Sue and Steve made me feel like part of the clan, and I enjoyed our times watching *America's Got Talent*, the Mariners, and hanging out with Linda, Roie, Gali, Mateo, and the rest of the family. They'll always be family to me.

Lady the cat, Sue and Steve's cat, was a good friend in this process as well.

My brothers, Barry (aka: Barrycuda) and Josh (aka: Young Squid), were always available to talk and kept me entertained amid long stretches of typing alone at the computer. They make life fun. Papoose for life.

The extended Kantrowitz and Stepner families are responsible for the person I am today. We owe it all to Leon and Miriam Kantrowitz and Jerome and Eleanor Stepner, who worked hard to ensure a future for us. And we continue to be inspired by my late cousin Rachael Kantrowitz, who showed us how to live a life filled with love and kindness. Thanks for everything, guys.

Carmel DeAmicis enthusiastically listened to me as I tried to make sense of my reporting. She read along and gave incredible advice every step of the way. Thanks Carmel, you're the best.

Mark Bergen, of *Bloomberg* and Bernal Heights fame, helped me game-plan the reporting for this book and has been an amazing, supportive friend. Our many bike rides through Sausalito and Funston, and various hikes throughout the Bay Area are the good stuff. Here's to many more.

Brad Allen kept me grounded through this process and has helped me look at life with a fresh perspective every time we talk. He is okay at basketball too.

Jessica Fraidlin showed me the ways of San Francisco, and I learned much about life, business, and food from her and her husband, Alex. They were always game to talk about the book's latest reporting challenges at our weekly dinners, and there to offer advice and support.

Jane Leibrock has been supportive and enthusiastic, cheering me on all the way. She's always been happy to listen and has never once told me to shut up. Which is proof she's a saint.

Nate Skid has inspired me to keep striving for bigger and better. His wife, Lang, and daughter, Evelyn, complete a family that inspires me.

Matt Sudol taught me to appreciate a moment in time. Richard Solomon taught me how advertising works. Howard Spieler continues to be my mentor despite no longer being compelled to do so by NYCEDC.

In the middle of this process, a bunch of friends showed up to my apartment for a "Book Draft Party," reading through my draft and giving feedback. Many of them are named here, some I can't name, and a special shout-out to Ariel Camus and Joe Wadlington, who kept us laughing throughout.

The North Shore crew, David, Gabe, Jenny, and Rebecca kept my spirits high throughout the process.

The Cornell crew, Ali, Ayala, Chad, Dan, Emily, Ezron, Gavi, Hannah, Herbie, Jack, Jasmine, Jaclyn, Josh, Judah, Lauren, Naomi, Newman, Nicole, Perry, Rachel, Rina, Ronit, Schapp, and Tzippy . . . you guys are amazing. Thanks for all the support and encouragement.

The Never Stop Never Stopping group chat kept me (somewhat) sane when it was just me and the keyboard on many days. They helped me learn with a steady stream of links and good discussion. Thanks, guys, for not kicking me out despite my minimal NBA knowledge.

Simon Dumenco, Michael Learmonth, Maureen Morrison, Matt Quinn, and Judy Pollack at *Ad Age* turned me from a marketing guy into a reporter.

Saul Austerlitz taught me about freelance journalism when I was just starting out, and then book writing when I wanted to jump into that too.

Larry Reibstein met with me in the early days and pointed me in the right direction. A small gesture made a big difference in my life.

Scott Olster gave me my big break, publishing my first story in *Fortune*, which led to the next, and the next after that.

Zack O'Malley Greenburg and Jon Bruner brought me into *Forbes* as a contributor, and that's how it all began.

Gary, Keith, and Ron of the #MetsBooth, thanks for keeping me company on many solitary afternoons throughout the spring and summer. Hope to see you guys in the fall sometime soon.

The fine folks at Arizmendi on Valencia always greeted me with a coffee and a smile on many, many visits there over the book writing process. No matter how deep into a writing hole I was, I knew I could stop by for thirty minutes of bliss every day.

The people who keep up Glen Canyon Park gave me a beautiful place to run nearly every day through this process. I recommend you visit the next time you're in San Francisco.

And to all those who told me I couldn't, wouldn't, or didn't have what it took—thank you. You lit my fire.

NOTES

PROLOGUE: THE ZUCKERBERG ENCOUNTER

2 **Zuckerberg had been hard at work on his "Manifesto":** Zuckerberg, Mark. "Building Global Community." Facebook, February 16, 2017. https://www.facebook.com/notes/mark-zuckerberg/building-global-community/10103508221158471.

INTRODUCTION: ALWAYS DAY ONE

5 **At an Amazon all hands in March 2017:** Amazon News. "Jeff Bezos on Why It's Always Day 1 at Amazon." YouTube, April 19, 2017. https://www.youtube.com/watch?v=fTwXS2H_iJo.

6 **By 2015, it was fifteen:** Lam, Bourree. "Where Do Firms Go When They Die?" *Atlantic.* Atlantic Media Company, April 12, 2015. https://www.theatlantic.com/business/archive/2015/04/where-do-firms-go-when-they-die/390249/.

12 **And after investors handed the company $225 million:** Winkler, Rolfe. "Software 'Robots' Power Surging Values for Three Little-Known Startups." *Wall Street Journal.* Dow Jones & Company, September 17,

2018. https://www.wsj.com/articles/software-robots-power-surging-values-for-three-little-known-startups-1537225425.

13 **UiPath's main competitors, Automation Anywhere, raised $300 million:** Lunden, Ingrid. "RPA Startup Automation Anywhere Nabs $300M from SoftBank at a $2.6B Valuation." TechCrunch. TechCrunch, November 15, 2018. https://techcrunch.com/2018/11/15/rpa-startup-automation-anywhere-nabs-300m-from-softbank-at-a-2-6b-valuation.

18 **Netflix, for instance, has a feedback culture:** Ramachandran, Shalini, and Joe Flint. "At Netflix, Radical Transparency and Blunt Firings Unsettle the Ranks." *Wall Street Journal.* Dow Jones & Company, October 25, 2018. https://www.wsj.com/articles/at-netflix-radical-transparency-and-blunt-firings-unsettle-the-ranks-1540497174?mod=hp_lead_pos4.

18 **Ideas at Tesla come from the top:** Duhigg, Charles. "Dr. Elon & Mr. Musk: Life Inside Tesla's Production Hell." *Wired.* Condé Nast, December 13, 2008. https://www.wired.com/story/elon-musk-tesla-life-inside-gigafactory.

18 **Uber's culture is famously troubled:** Isaac, Mike. *Super Pumped: The Battle for Uber.* New York: W. W. Norton & Company, 2019.

CHAPTER 1: INSIDE JEFF BEZOS'S CULTURE OF INVENTION

23 **Bezos drives Amazon's inventive culture through fourteen leadership principles:** "Leadership Principles." Amazon.jobs. Accessed October 3, 2019. https://www.amazon.jobs/en/principles.

26 **"No powerpoint presentations from now on," he wrote:** Stone, Madeline. "A 2004 Email from Jeff Bezos Explains Why PowerPoint Presentations Aren't Allowed at Amazon." *Business Insider.* Business Insider, July 28, 2015. https://www.businessinsider.com/jeff-bezos-email-against-powerpoint-presentations-2015-7.

27 **The memo was exhaustive:** These memos even have their own set of micro-leadership principles for each group, called tenets.

31 **In March 2012, Amazon acquired Kiva Systems:** Rusli, Evelyn. "Amazon.com to Acquire Manufacturer of Robotics." *New York Times.* New York Times, March 19, 2012. https://dealbook.nytimes.com/2012/03/19/amazon-com-buys-kiva-systems-for-775-million/.

31 **Amazon placed roughly 15,000 robots in its FCs by 2014:** Seetharaman, Deepa. "Amazon Has Installed 15,000 Warehouse Robots to Deal with Increased Holiday Demand." *Business Insider.* Business Insider, December 1, 2014. https://www.businessinsider.com/r-amazon-rolls-out-kiva-robots-for-holiday-season-onslaught-2014-12.

31 **and had 30,000 in operation by 2015:** Levy, Nat. "Chart: Amazon Robots on the Rise, Gaining Slowly but Steadily on Human Workforce." *GeekWire.* GeekWire, December 29, 2016. https://www.geekwire.com/2016/chart-amazon-robots-rise-gaining-slowly-steadily-human-workforce/.

32 **Amazon seems likely to automate other core parts of FC work:** Del Rey, Jason. "Land of the Giants." *Vox.* Accessed October 3, 2019. https://www.vox.com/land-of-the-giants-podcast.

33 **they didn't want to take a bathroom break:** Pollard, Chris. "Rushed Amazon Staff Pee into Bottles as They're Afraid of Time-Wasting." *Sun.* Sun, April 15, 2018. https://www.thesun.co.uk/news/6055021/rushed-amazon-warehouse-staff-time-wasting.

33 **The company's corporate staff:** Stone, Brad. *The Everything Store: Jeff Bezos and the Age of Amazon.* New York: Little, Brown and Company, 2013.

35 **"Because of the challenges":** Recode. "Amazon Employee Work-Life Balance | Jeff Bezos, CEO Amazon | Code Conference 2016." YouTube, June 2, 2016. https://www.youtube.com/watch?v=PTYFEgXaRbU.

47 **"Customers are always unsatisfied," Bezos said:** TheBushCenter. "Forum on Leadership: A Conversation with Jeff Bezos." YouTube, April 20, 2018. https://www.youtube.com/watch?v=xu6vFIKAUxk.

48 **a brutal five-thousand-word *New York Times* article:** Kantor, Jodi, and David Streitfeld. "Inside Amazon: Wrestling Big Ideas in a Bruising Workplace." *New York Times.* New York Times, August 15, 2015. https://www.nytimes.com/2015/08/16/technology/inside-amazon-wrestling-big-ideas-in-a-bruising-workplace.html.

49 **Amazon went to war with the *New York Times*:** Carney, Jay. "What the New York Times Didn't Tell You." Medium. Medium, October 19, 2015. https://medium.com/@jaycarney/what-the-new-york-times-didn-t-tell-you-a1128aa78931.

49 ***New York Times* editor Dean Baquet shot back:** Communications, NYTCo. "Dean Baquet Responds to Jay Carney's Medium Post." Medium. Medium, October 19, 2015. https://medium.com/@NYTimesComm/dean-baquet-responds-to-jay-carney-s-medium-post-6af794c7a7c6.

49 **When the article hit, Bezos emailed the company:** Cook, John. "Full Memo: Jeff Bezos Responds to Brutal NYT Story, Says It Doesn't Represent the Amazon He Leads." *GeekWire.* GeekWire, August 16, 2015. https://www.geekwire.com/2015/full-memo-jeff-bezos-responds-to-cutting-nyt-expose-says-tolerance-for-lack-of-empathy-needs-to-be-zero/.

CHAPTER 2: INSIDE MARK ZUCKERBERG'S CULTURE OF FEEDBACK

64 **Sandberg's conference room, Only Good News:** Inskeep, Steve. "We Did Not Do Enough to Protect User Data, Facebook's Sandberg Says." NPR. NPR, April 6, 2018. https://www.npr.org/2018/04/06/600071401/we-did-not-do-enough-to-protect-user-data-facebooks-sandberg-says.

67 **Facebook released a native iOS app:** Rusli, Evelyn M. "Even Facebook Must Change." *Wall Street Journal.* Dow Jones & Company, January 29, 2013. https://www.wsj.com/articles/SB10001424127887323829504578272233666653120.

68 **today more than 90 percent of Facebook's advertising revenue comes from mobile:** Goode, Lauren. "Facebook Was Late to Mobile. Now Mobile Is the Future." *Wired.* Condé Nast, February 06, 2019. https://www.wired.com/story/facebooks-future-is-mobile/.

69 **people were sharing fewer original posts:** Efrati, Amir. "Facebook Struggles to Stop Decline in 'Original' Sharing." *The Information,* April 7, 2016. https://www.theinformation.com/articles/facebook-struggles-to-stop-decline-in-original-sharing?shared=5dd15d.

69 **its network was now more than 1.5 billion users:** Facebook 10-Q. Accessed October 3, 2019. https://www.sec.gov/Archives/edgar/data/1326801/000132680115000032/fb-9302015x10q.htm.

70 **With Groups membership climbing by tens of millions:** Kantrowitz, Alex. "Small Social Is Here: Why Groups Are Finally Finding a

Home Online." *BuzzFeed News*. BuzzFeed News, November 3, 2015. https://www.buzzfeednews.com/article/alexkantrowitz/small-social-is-here-why-groups-are-finally-finding-a-home-o.

73 **called "Project Voldemort":** Wells, Georgia, and Deepa Seetharaman. "WSJ News Exclusive | Snap Detailed Facebook's Aggressive Tactics in 'Project Voldemort' Dossier." *Wall Street Journal*. Dow Jones & Company, September 24, 2019. https://www.wsj.com/articles/snap-detailed-facebooks-aggressive-tactics-in-project-voldemort-dossier-11569236404.

75 **Six months later, Facebook bought Face.com:** Tsotsis, Alexia. "Facebook Scoops Up Face.com for $55–60M to Bolster Its Facial Recognition Tech (Updated)." *TechCrunch*. TechCrunch, June 18, 2012. https://techcrunch.com/2012/06/18/facebook-scoops-up-face-com-for-100m-to-bolster-its-facial-recognition-tech/.

77 **Zuckerberg's product team rolled out Live:** Kantrowitz, Alex. "Facebook Expands Live Video Beyond Celebrities." *BuzzFeed News*. BuzzFeed News, December 3, 2015. https://www.buzzfeednews.com/article/alexkantrowitz/facebook-brings-its-live-streaming-to-the-masses#.jegRRDmJK.

77 **Donesha Gantt went live on Facebook after being shot:** Rabin, Charles. "Woman Posts Live Video of Herself After Being Shot in Opa-Locka Burger King Drive-Through." *Miami Herald*. Miami Herald, February 2, 2016. https://www.miamiherald.com/news/local/crime/article57897483.html.

77 **aired videos of graphic violence at a rate of about twice per month:** Kantrowitz, Alex. "Violence on Facebook Live Is Worse Than You Thought." *BuzzFeed News*. BuzzFeed News, June 16, 2017. https://www.buzzfeednews.com/article/alexkantrowitz/heres-how-bad-facebook-lives-violence-problem-is.

80 **rolling out an AI-based suicide-prevention tool:** Kantrowitz, Alex. "Facebook Is Using Artificial Intelligence to Help Prevent Suicide." *BuzzFeed News*. BuzzFeed News, March 1, 2017. https://www.buzzfeednews.com/article/alexkantrowitz/facebook-is-using-artificial-intelligence-to-prevent-suicide.

80 **an update on how the overall program was performing:** Rosen, Guy.

"F8 2018: Using Technology to Remove the Bad Stuff Before It's Even Reported." Facebook Newsroom, May 2, 2018. https://newsroom.fb.com/news/2018/05/removing-content-using-ai/.

80 **some of its moderators work in miserable conditions:** Newton, Casey. "The Secret Lives of Facebook Moderators in America." *Verge*. Vox, February 25, 2019. https://www.theverge.com/2019/2/25/18229714/cognizant-facebook-content-moderator-interviews-trauma-working-conditions-arizona.

83 **a large-scale Kremlin-sponsored misinformation campaign:** Stamos, Alex. "An Update on Information Operations on Facebook." Facebook Newsroom, September 6, 2017. https://newsroom.fb.com/news/2017/09/information-operations-update/.

83 **Cambridge Analytica, a data analytics firm, illicitly used:** Rosenberg, Matthew, Nicholas Confessore, and Carole Cadwalladr. "How Trump Consultants Exploited the Facebook Data of Millions." *New York Times*. New York Times, March 17, 2018. https://www.nytimes.com/2018/03/17/us/politics/cambridge-analytica-trump-campaign.html.

84 ***best in* "The Ugly":** Mac, Ryan, Charlie Warzel, and Alex Kantrowitz. "Growth at Any Cost: Top Facebook Executive Defended Data Collection in 2016 Memo—and Warned That Facebook Could Get People Killed." *BuzzFeed News*. BuzzFeed News, March 29, 2018. https://www.buzzfeednews.com/article/ryanmac/growth-at-any-cost-top-facebook-executive-defended-data.

86 **his opening statement:** Stewart, Emily. "What Mark Zuckerberg Will Tell Congress About the Facebook Scandals." *Vox*. Vox, April 10, 2018. https://www.vox.com/policy-and-politics/2018/4/9/17215640/mark-zuckerberg-congress-testimony-facebook.

89 **Cameroon:** McAllister, Edward. "Facebook's Cameroon Problem: Stop Online Hate Stoking Conflict." Reuters. Thomson Reuters, November 4, 2018. https://www.reuters.com/article/us-facebook-cameroon-insight/facebooks-cameroon-problem-stop-online-hate-stoking-conflict-idUSKCN1NA0GW.

89 **and Sri Lanka:** Rajagopalan, Megha. "'We Had to Stop Facebook': When Anti-Muslim Violence Goes Viral." *BuzzFeed News*. BuzzFeed News, April 7, 2018. https://www.buzzfeednews.com/article/meghara/we-had-to-stop-facebook-when-anti-muslim-violence-goes-viral.

CHAPTER 3: INSIDE SUNDAR PICHAI'S CULTURE OF COLLABORATION

94 **As the debate raged:** Conger, Kate. "Exclusive: Here's the Full 10-Page Anti-Diversity Screed Circulating Internally at Google [Updated]." *Gizmodo*. Gizmodo, August 5, 2017. https://gizmodo.com/exclusive-heres-the-full-10-page-anti-diversity-screed-1797564320.

94 **the Me Too movement:** Alyssa Milano, Twitter Post, October 15, 2017, 1:21 p.m., https://twitter.com/Alyssa_Milano/status/919659438700670976.

99 **When Marissa Mayer:** Harmanci, Reyhan. "Inside Google's Internal Meme Generator." *BuzzFeed News*. BuzzFeed News, September 26, 2012. https://www.buzzfeednews.com/article/reyhan/inside-googles-internal-meme-generator.

100 **65 percent of Google's search traffic:** Nelson, Jeff. "What Did Sundar Pichai Do That His Peers Didn't, to Get Promoted Through the Ranks from an Entry Level PM to CEO of Google?" Quora, July 24, 2016. https://www.quora.com/What-did-Sundar-Pichai-do-that-his-peers-didnt-to-get-promoted-through-the-ranks-from-an-entry-level-PM-to-CEO-of-Google/answer/Jeff-Nelson-32?ch=10&share=53473102&srid=au3.

102 **"what do you think of Gmail":** "Sundar Pichai Full Speech at IIT Kharagpur 2017 | Sundar Pichai at KGP | Latest Speech." YouTube, January 10, 2017. https://www.youtube.com/watch?v=-yLlMk41sro&feature=youtu.be.

103 **Google acquired Upstartle:** Mazzon, Jen. "Writely So." *Official Google Blog*, March 9, 2006. https://googleblog.blogspot.com/2006/03/writely-so.html.

103 **it introduced Google Calendar:** Sjogreen, Carl. "It's About Time." Official Google Blog, April 13, 2006. https://googleblog.blogspot.com/2006/04/its-about-time.html.

103 **it introduced Google Spreadsheets:** Rochelle, Jonathan. "It's Nice to Share." Official Google Blog, June 6, 2006. https://googleblog.blogspot.com/2006/06/its-nice-to-share.html.

106 **Pichai said as he introduced Chrome:** "Sundar Pichai Launching Google Chrome." YouTube, February 19, 2017. https://www.youtube.com/watch?v=3_Ye38fBQMo.

107 **Chrome debuted in 2008:** Doerr, John E. *Measure What Matters: How Google, Bono, and the Gates Foundation Rock the World with OKRs.* New York: Portfolio, 2018.

107 **ceased developing Internet Explorer:** Newcomb, Alyssa. "Microsoft: Drag Internet Explorer to the Trash. No, Really." *Fortune*. Fortune, February 8, 2019. https://fortune.com/2019/02/08/download-internet-explorer-11-end-of-life-microsoft-edge/?xid=gn_editorspicks.

109 **releasing the Amazon Echo and its embedded digital assistant:** Stone, Brad, and Spencer Soper. "Amazon Unveils a Listening, Talking, Music-Playing Speaker for Your Home." *Bloomberg*. Bloomberg, November 6, 2014. https://www.bloomberg.com/news/articles/2014-11-06/amazon-echo-is-a-listening-talking-music-playing-speaker-for-your-home.

110 **Larry Page published a shocking blog post:** Page, Larry. "G Is for Google." Official Google Blog, August 10, 2015. https://googleblog.blogspot.com/2015/08/google-alphabet.html.

111 **apps would account for 89.2 percent of all time spent:** "US Time Spent with Media: EMarketer's Updated Estimates and Forecast for 2014–2019." eMarketer, April 27, 2017. https://www.emarketer.com/Report/US-Time-Spent-with-Media-eMarketers-Updated-Estimates-Forecast-20142019/2002021.

116 **there isn't much room for listening:** Pierce, David. "One Man's Quest to Make Google's Gadgets Great." *Wired*. Condé Nast, February 8, 2018. https://www.wired.com/story/one-mans-quest-to-make-googles-gadgets-great/.

119 **Liz Fong-Jones:** Tiku, Nitasha. "Three Years of Misery Inside Google, the Happiest Company in Tech." *Wired*. Condé Nast, August 13, 2019. https://www.wired.com/story/inside-google-three-years-misery-happiest-company-tech/.

119 **the Googlers wrote a protest letter:** Shane, Scott, and Daisuke Wakabayashi. " 'The Business of War': Google Employees Protest Work for the Pentagon." *New York Times*. New York Times, April 4, 2018. https://www.nytimes.com/2018/04/04/technology/google-letter-ceo-pentagon-project.html?login=smartlock&auth=login-smartlock.

120 **an international letter against the use of AI for autonomous warfare:** "Lethal Autonomous Weapons Pledge." Future of Life Institute. https://futureoflife.org/lethal-autonomous-weapons-pledge/.

120 **"Hey, I left the Defense Department":** Tarnoff, Ben. "Tech Workers Versus the Pentagon." *Jacobin*. Jacobin, June 6, 2018. https://jacobinmag.com/2018/06/google-project-maven-military-tech-workers.

120 **about a dozen Googlers resigned:** Conger, Kate. "Google Employees Resign in Protest Against Pentagon Contract." *Gizmodo*. Gizmodo, May 14, 2018. https://gizmodo.com/google-employees-resign-in-protest-against-pentagon-con-1825729300.

120 **an ensuing leak:** Shane, Scott, Cade Metz, and Daisuke Wakabayashi. "How a Pentagon Contract Became an Identity Crisis for Google." *New York Times*. New York Times, May 30, 2018. https://www.nytimes.com/2018/05/30/technology/google-project-maven-pentagon.html.

122 **Pichai released the AI Principles:** Pichai, Sundar. "AI at Google: Our Principles." Google, June 7, 2018. https://www.blog.google/technology/ai/ai-principles/.

122 **Google said it would not renew:** Alba, Davey. "Google Backs Away from Controversial Military Drone Project." *BuzzFeed News*. BuzzFeed News, June 1, 2018. https://www.buzzfeednews.com/article/daveyalba/google-says-it-will-not-follow-through-on-pentagon-drone-ai.

123 **The Walkout, as it's now known:** Wakabayashi, Daisuke, and Katie Benner. "How Google Protected Andy Rubin, the 'Father of Android'." *New York Times*. New York Times, October 25, 2018. https://www.nytimes.com/2018/10/25/technology/google-sexual-harassment-andy-rubin.html.

124 **In an email to the moms group:** Morris, Alex. "Rage Drove the Google Walkout. Can It Bring About Real Change at Tech Companies?" *New York*. New York Magazine, February 5, 2019. http://nymag.com/intelligencer/2019/02/can-the-google-walkout-bring-about-change-at-tech-companies.html.

126 **"Some of you have raised very constructive ideas":** Fried, Ina. "Google CEO: Apology for Past Harassment Issues Not Enough." *Axios*. Axios, October 30, 2018. https://www.axios.com/google-ceo-apologizes-past-sexual-harassment-aec53899-6ac0-4a70-828d-70c263e56305.html.

126 **"rolling thunder" of action:** Ghaffary, Shirin, and Eric Johnson. "After 20,000 Workers Walked Out, Google Said It Got the Message. The Workers Disagree." *Vox*. Vox, November 21, 2018. https://www.vox

.com/2018/11/21/18105719/google-walkout-real-change-organizers-protest-discrimination-kara-swisher-recode-decode-podcast.

127 **the end of forced arbitration:** Wakabayashi, Daisuke. "Google Ends Forced Arbitration for All Employee Disputes." *New York Times*. New York Times, February 21, 2019. https://www.nytimes.com/2019/02/21/technology/google-forced-arbitration.html.

127 **Google also retaliated:** Tiku, Nitasha. "Google Walkout Organizers Say They're Facing Retaliation." *Wired*. Condé Nast, April 22, 2019. https://www.wired.com/story/google-walkout-organizers-say-theyre-facing-retaliation/.

128 **confidence in Pichai and his leadership team dropped:** Kowitt, Beth. "Inside Google's Civil War." *Fortune*. Fortune, May 17, 2019. https://fortune.com/longform/inside-googles-civil-war/.

CHAPTER 4: TIM COOK AND THE APPLE QUESTION

129 **All this makes Brownlee's review:** Brownlee, Marques. "Apple HomePod Review: The Dumbest Smart Speaker?" YouTube, February 16, 2018. https://www.youtube.com/watch?v=mpjREfvZiDs&feature=youtu.be.

134 **Angela Ahrendts, the former Burberry:** Gruber, John. "Angela Ahrendts to Leave Apple in April; Deirdre O'Brien Named Senior Vice President of Retail and People." *Daring Fireball* (blog). Accessed February 5, 2019. https://daringfireball.net/linked/2019/02/05/ahrendts-obrien.

134 **As would Jony Ive:** Gruber, John. "Jony Ive Is Leaving Apple." *Daring Fireball* (blog), June 27, 2019. https://daringfireball.net/2019/06/jony_ive_leaves_apple.

135 **a leaked United Airlines document:** Mayo, Benjamin. "United Airlines Takes Down Poster That Revealed Apple Is Its Largest Corporate Spender." *9to5Mac*, January 14, 2019. https://9to5mac.com/2019/01/14/united-airlines-apple-biggest-customer/.

139 **It was a costly mistake:** Schleifer, Theodore. "An Apple Engineer Showed His Daughter the New IPhone X. Now, She Says, He's Fired." *Recode*. Vox, October 29, 2017. https://www.vox.com/2017/10/29/16567244/apple-engineer-fired-iphone-x-daughter-secret-product-launch.

140 **a rare letter:** Cook, Tim. "Letter from Tim Cook to Apple Inves-

tors." Apple Newsroom, January 2, 2019. https://www.apple.com/newsroom/2019/01/letter-from-tim-cook-to-apple-investors/.

140 **revised its financial predictions:** Thompson, Ben. "Apple's Errors." *Stratechery by Ben Thompson*, January 7, 2019. https://stratechery.com/2019/apples-errors/?utm_source=Memberful&utm_campaign=131ddd5a64-weekly_article_2019_01_07&utm_medium=email&utm_term=0_d4c7fece27-131ddd5a64-110945413.

141 **"I'm happy with my iPhone 8":** Balakrishnan, Anita, and Deirdre Bosa. "Apple Co-Founder Steve Wozniak: iPhone X Is the First iPhone I Won't Buy on 'Day One.'" *CNBC*. CNBC, October 23, 2017. https://www.cnbc.com/2017/10/23/apple-co-founder-steve-wozniak-not-upgrading-to-iphone-x-right-away.html.

141 **Cook, in an interview with CNBC:** "CNBC Exclusive: CNBC Transcript: Apple CEO Tim Cook Speaks with CNBC's Jim Cramer Today." *CNBC*. CNBC, January 8, 2019. https://www.cnbc.com/2019/01/08/exclusive-cnbc-transcript-apple-ceo-tim-cook-speaks-with-cnbcs-jim-cramer-today.html.

142 **Apple had Siri:** Gross, Doug. "Apple Introduces Siri, Web Freaks Out." *CNN*. Cable News Network, October 4, 2011. https://www.cnn.com/2011/10/04/tech/mobile/siri-iphone-4s-skynet/index.html.

143 **"Since October 2011, when Steve died":** Note that Jobs began the Siri project.

147 **push the release back:** Hall, Zac. "Apple Delaying HomePod Smart Speaker Launch until next Year." *9to5Mac*, November 17, 2017. https://9to5mac.com/2017/11/17/homepad-delay/.

151 **Apple moved two hundred employees off its struggling Project Titan:** Kolodny, Lora, Christina Farr, and Paul A. Eisenstein. "Apple Just Dismissed More than 200 Employees from Project Titan, Its Autonomous Vehicle Group." *CNBC*. CNBC, January 24, 2019. https://www.cnbc.com/2019/01/24/apple-lays-off-over-200-from-project-titan-autonomous-vehicle-group.html.

153 **"How is the work culture":** "How Is the Work Culture at the IS&T Division of Apple?" Quora. https://www.quora.com/How-is-the-work-culture-at-the-IS-T-division-of-Apple.

155 **fifteen-dollar-per-hour wage floor:** Salinas, Sara. "Amazon Raises Minimum Wage to $15 for All US Employees." *CNBC*. CNBC, October 2,

2018. https://www.cnbc.com/2018/10/02/amazon-raises-minimum-wage-to-15-for-all-us-employees.html.

155 **$28,000 per year:** Gross, Terry. "For Facebook Content Moderators, Traumatizing Material Is a Job Hazard." *NPR*. NPR, July 1, 2019. https://www.npr.org/2019/07/01/737498507/for-facebook-content-moderators-traumatizing-material-is-a-job-hazard.

155 **San Bernardino, California:** Nagourney, Adam, Ian Lovett, and Richard Pérez-Peña. "San Bernardino Shooting Kills at Least 14; Two Suspects Are Dead." *New York Times*. New York Times, December 2, 2015. https://www.nytimes.com/2015/12/03/us/san-bernardino-shooting.html.

155 **an iPhone 5c:** Ng, Alfred. "FBI Asked Apple to Unlock iPhone Before Trying All Its Options." CNET, March 27, 2018. https://www.cnet.com/news/fbi-asked-apple-to-unlock-iphone-before-trying-all-its-options.

156 **not just one iPhone:** Grossman, Lev. "Apple CEO Tim Cook: Inside His Fight with the FBI." *Time*. Time Magazine, March 17, 2016. https://time.com/4262480/tim-cook-apple-fbi-2.

156 **the side of privacy:** Cook, Tim. "Customer Letter." Apple. Accessed February 16, 2016. https://www.apple.com/customer-letter.

159 **"To me, marketing is about values":** "Best Marketing Strategy Ever! Steve Jobs Think Different / Crazy Ones Speech (with Real Subtitles)." YouTube, April 21, 2013. https://www.youtube.com/watch?v=keCwRdbwNQY.

161 **as Oprah put it:** Albergotti, Reed. "Apple's 'Show Time' Event Puts the Spotlight on Subscription Services." *Washington Post*. Washington Post, March 25, 2019. https://www.washingtonpost.com/technology/2019/03/25/apple-march-event-streaming-news-subscription.

CHAPTER 5: SATYA NADELLA AND THE MICROSOFT CASE STUDY

165 **a former aQuantive manager told *GeekWire*:** Cook, John. "After the Writedown: How Microsoft Squandered Its $6.3B Buy of Ad Giant aQuantive." *GeekWire*. GeekWire, July 12, 2012. https://www.geekwire.com/2012/writedown-microsoft-squandered-62b-purchase-ad-giant-aquantive/.

165 **an advance copy:** Bishop, Todd. "Microsoft's 'Lost Decade'? Vanity Fair Piece Is Epic, Accurate and Not Entirely Fair." *GeekWire*. GeekWire, July 4, 2012. https://www.geekwire.com/2012/microsofts-lost-decade-vanity-fair-piece-accurate-incomplete.

165 **"What began as a lean competition machine":** Eichenwald, Kurt. "How Microsoft Lost Its Mojo: Steve Ballmer and Corporate America's Most Spectacular Decline." *Vanity Fair*. Vanity Fair, July 24, 2012. https://www.vanityfair.com/news/business/2012/08/microsoft-lost-mojo-steve-ballmer.

165 **Ballmer stepped down:** Bishop, Todd. "Microsoft Names Satya Nadella CEO; Bill Gates Stepping Down as Chairman to Serve as Tech Adviser." *GeekWire*. GeekWire, February 4, 2014. https://www.geekwire.com/2014/microsoft-ceo-main.

167 **a $13 billion business:** Fontana, John. "Microsoft Tops $60 Billion in Annual Revenue." Network World, July 17, 2008. https://www.networkworld.com/article/2274218/microsoft-tops--60-billion-in-annual-revenue.html.

167 **20 percent of Microsoft's total revenue:** Romano, Benjamin. "Microsoft Server and Tools Boss Muglia Given President Title." *Seattle Times*. Seattle Times Company, January 6, 2009. https://www.seattletimes.com/business/microsoft/microsoft-server-and-tools-boss-muglia-given-president-title.

169 **AWS controlled 37 percent:** D'Onfro, Jillian. "Here's a Reminder Just How Massive Amazon's Web Services Business Is." *Business Insider*. Business Insider, June 16, 2014. https://www.businessinsider.com/amazon-web-services-market-share-2014-6.

170 **Ballmer promoted Satya Nadella:** Foley, Mary Jo. "Meet Microsoft's New Server and Tools Boss: Satya Nadella." *ZDNet*, February 9, 2011. https://www.zdnet.com/article/meet-microsofts-new-server-and-tools-boss-satya-nadella.

171 **His final major act:** Warren, Tom. "Microsoft Writes Off $7.6 Billion from Nokia Deal, Announces 7,800 Job Cuts." *Verge*. Vox, July 8, 2015. https://www.theverge.com/2015/7/8/8910999/microsoft-job-cuts-2015-nokia-write-off.

172 **In an email to employees:** "Satya Nadella Email to Employees on First Day as CEO." Microsoft News Center, February 4, 2014. https://news

.microsoft.com/2014/02/04/satya-nadella-email-to-employees-on-first-day-as-ceo.

173 **as much startup thinking as possible:** Nadella, Satya. *Hit Refresh: The Quest to Rediscover Microsoft's Soul and Imagine a Better Future for Everyone.* New York: HarperCollins, 2017.

173 **Nadella also expanded Microsoft Garage:** Choney, Suzanne. "Microsoft Garage Expands to Include Exploration, Creation of Cross-Platform Consumer Apps." *Fire Hose* (blog), October 22, 2014. https://web.archive.org/web/20141025020143/http://blogs.microsoft.com/firehose/2014/10/22/microsoft-garage-expands-to-include-exploration-creation-of-cross-platform-consumer-apps.

176 **Nadella said at the time:** Lunden, Ingrid. "Microsoft Forms New AI Research Group Led by Harry Shum." *TechCrunch.* TechCrunch, September 29, 2016. https://techcrunch.com/2016/09/29/microsoft-forms-new-ai-research-group-led-by-harry-shum.

181 **one you can still watch on YouTube:** MasterBlackHat. "Steve Ballmer—Dance Monkey Boy!" YouTube, December 28, 2007. https://www.youtube.com/watch?v=edN4o8F9_P4.

183 **Microsoft is a company of conflicting interests:** Cornet, Manu. "Organizational Charts." Accessed October 7, 2019. http://bonkersworld.net/organizational-charts.

185 **In *Mindset*, her 2007 book:** Dweck, Carol S. *Mindset: The New Psychology of Success.* New York: Random House, 2007.

185 **"We need to be open to the ideas of others":** Bishop, Todd. "Exclusive: Satya Nadella Reveals Microsoft's New Mission Statement, Sees 'Tough Choices' Ahead." *GeekWire.* GeekWire, June 25, 2015. https://www.geekwire.com/2015/exclusive-satya-nadella-reveals-microsofts-new-mission-statement-sees-more-tough-choices-ahead.

185 **Nadella drove this point home by demoing Office for iOS:** Kim, Eugene. "Microsoft CEO Satya Nadella Just Used an iPhone to Demo Outlook." *Business Insider.* Business Insider, September 16, 2015. https://www.businessinsider.com/microsoft-ceo-satya-nadella-used-iphone-2015-9.

186 **Microsoft's "Biggest Reorganization in Years":** Bass, Dina, and Ian King. "Microsoft Unveils Biggest Reorganization in Years." *Bloomberg.*

Bloomberg, March 29, 2018. https://www.bloomberg.com/news/articles/2018-03-29/microsoft-unveils-biggest-reorganization-in-years-as-myerson-out.

186 **"We can't let any organizational boundaries get in the way":** Nadella, Satya. "Satya Nadella Email to Employees: Embracing Our Future: Intelligent Cloud and Intelligent Edge." Microsoft News Center, March 29, 2018. https://news.microsoft.com/2018/03/29/satya-nadella-email-to-employees-embracing-our-future-intelligent-cloud-and-intelligent-edge.

186 **When Microsoft acquired LinkedIn:** Lunden, Ingrid. "Microsoft Officially Closes Its $26.2B Acquisition of LinkedIn." *TechCrunch*. TechCrunch, December 8, 2016. https://techcrunch.com/2016/12/08/microsoft-officially-closes-its-26-2b-acquisition-of-linkedin/.

187 **LinkedIn's revenue:** Warren, Tom. "Microsoft's Bets on Surface, Gaming, and LinkedIn Are Starting to Pay Off." *Verge*. Vox, April 26, 2018. https://www.theverge.com/2018/4/26/17286900/microsoft-q3-2018-earnings-cloud-surface-linkedin-revenue.

188 **Nadella was on the email chain:** Gershgorn, Dave. "Amid Employee Uproar, Microsoft Is Investigating Sexual Harassment Claims Overlooked by HR." *Quartz*. Quartz, April 4, 2019. https://qz.com/1587477/microsoft-investigating-sexual-harassment-claims-overlooked-by-hr/.

190 **Today the company's market cap is more than $1 trillion:** $1 trillion market cap as of October 2019.

CHAPTER 6: A LOOK INTO THE BLACK MIRROR

191 **The science fiction series *Black Mirror* debuted:** Brooker, Charlie. "Charlie Brooker: The Dark Side of Our Gadget Addiction." *Guardian*. Guardian, December 1, 2011. https://www.theguardian.com/technology/2011/dec/01/charlie-brooker-dark-side-gadget-addiction-black-mirror.

192 **"I'm a natural worrier":** "Charlie Brooker on Black Mirror vs Reality | Good Morning Britain." *Good Morning Britain*. YouTube, October 30, 2018. https://www.youtube.com/watch?v=Na-ZIwy1bNI.

192 **China has implemented a social rating system:** Bruney, Gabrielle. "A 'Black Mirror' Episode Is Coming to Life in China." *Esquire*. Esquire,

March 17, 2018. https://www.esquire.com/news-politics/a19467976/black-mirror-social-credit-china.

192 **"a private part of his anatomy" into a dead pig's mouth:** Ashcroft, Michael, and Isabel Oakeshott. "David Cameron, Drugs, Debauchery and the Making of an Extraordinary Prime Minister." *Daily Mail Online.* Associated Newspapers, September 21, 2015. https://www.dailymail.co.uk/news/article-3242504/Drugs-debauchery-making-extraordinary-Prime-Minister-years-rumours-dogged-truth-shockingly-decadent-Oxford-days-gifted-Bullingdon-boy.html.

195 **ousted from the New America Foundation:** Vogel, Kenneth P. "Google Critic Ousted from Think Tank Funded by the Tech Giant." *New York Times.* New York Times, August 30, 2017. https://www.nytimes.com/2017/08/30/us/politics/eric-schmidt-google-new-america.html.

195 **Newspaper advertising revenue, meanwhile, dropped:** "Trends and Facts on Newspapers: State of the News Media." Pew Research Center's Journalism Project. Pew Research Center, July 9, 2019. https://www.journalism.org/fact-sheet/newspapers.

195 **newsroom employment at US newspapers dropped 47 percent:** Grieco, Elizabeth. "U.S. Newsroom Employment Has Dropped a Quarter since 2008, with Greatest Decline at Newspapers." Pew Research Center, July 9, 2019. https://www.pewresearch.org/fact-tank/2019/07/09/u-s-newsroom-employment-has-dropped-by-a-quarter-since-2008.

197 **153 artificial intelligence professors have left:** Metz, Cade. "When the A.I. Professor Leaves, Students Suffer, Study Says." *New York Times.* New York Times, September 6, 2019. https://www.nytimes.com/2019/09/06/technology/when-the-ai-professor-leaves-students-suffer-study-says.html.

199 **"The nuclear family structure is in statistical collapse":** Matheson, Boyd. "Why Do We Hate Each Other? A Conversation with Nebraska Sen. Ben Sasse (Podcast)." *Deseret News.* Deseret News, October 17, 2018. https://www.deseret.com/2018/10/17/20656288/why-do-we-hate-each-other-a-conversation-with-nebraska-sen-ben-sasse-podcast.

199 **Sasse calls loneliness an "epidemic":** Sasse, Ben. *Them: Why We Hate Each Other—and How to Heal.* New York: St. Martin's Press, 2018.

199 **"We become accustomed to the constant social stimulation":** Turkle, Sherry. *Alone Together: Why We Expect More from Technology and Less from Each Other.* New York: Basic Books, 2012.

199 **study on loneliness:** "Cigna U.S. Loneliness Index." Cigna, May, 2018, https://www.multivu.com/players/English/8294451-cigna-us-loneliness-survey/docs/IndexReport_1524069371598-173525450.pdf.

200 **"They have no religion":** Ravitz, Jessica. "Is the Internet Killing Religion?" *CNN.* CNN, April 9, 2014. http://religion.blogs.cnn.com/2014/04/09/is-the-internet-killing-religion/comment-page-6/.

200 **"Internet use decreases the chance of religious affiliation":** Downey, Allen. "Religious Affiliation, Education and Internet Use." *Religious Affiliation, Education and Internet Use*, 2014.

200 **23 percent of the population:** Shermer, Michael. "The Number of Americans with No Religious Affiliation Is Rising." *Scientific American.* Scientific American, April 1, 2018. https://www.scientificamerican.com/article/the-number-of-americans-with-no-religious-affiliation-is-rising.

200 **two hundred million people:** Kastrenakes, Jacob. "Facebook Adds New Group Tools as It Looks for 'Meaningful' Conversations." *Verge.* Vox, February 7, 2019. https://www.theverge.com/2019/2/7/18215564/facebook-groups-new-community-tools-mentorship.

201 **one billion by 2022:** Ortutay, Barbara. "Facebook Wants to Nudge You into 'Meaningful' Online Groups." *AP News.* Associated Press, June 22, 2017. https://www.apnews.com/713f8f66e88b45828fd62b1693652ee7.

201 **"Zuckerberg recognizes something that so few Christian leaders have":** Syverson, Andrea. "Commentary: Can Facebook Replace Churches?" *Salt Lake Tribune.* Salt Lake Tribune, July 6, 2017. https://archive.sltrib.com/article.php?id=5479818&itype=CMSID.

201 **life expectancy in the US:** Kight, Stef W. "Life Expectancy Drops in the U.S. for Third Year in a Row." *Axios.* Axios, November 29, 2018. https://www.axios.com/united-states-life-expectancy-drops-6881f610-3ca0-4758-b637-dd9c02b237d0.html.

201 **people died from drug overdoses:** "Drug and Opioid-Involved Overdose Deaths—United States, 2013–2017 | MMWR." Centers for Disease Control and Prevention, January 4, 2019. https://www.cdc.gov/mmwr/volumes/67/wr/mm675152e1.htm.

201 **Americans died by suicide:** Godlasky, Anne, and Alia E. Dastagir. "Suicide Rate up 33% in Less than 20 Years, Yet Funding Lags Behind Other Top Killers." *USA Today*. Gannett Satellite Information Network, December 21, 2018. https://www.usatoday.com/in-depth/news/investigations/surviving-suicide/2018/11/28/suicide-prevention-suicidal-thoughts-research-funding/971336002.

201 **"There's not a part of the country that has not been touched by this":** Boddy, Jessica. "The Forces Driving Middle-Aged White People's 'Deaths of Despair.'" *NPR*. NPR, March 23, 2017. https://www.npr.org/sections/health-shots/2017/03/23/521083335/the-forces-driving-middle-aged-white-peoples-deaths-of-despair.

201 **unemployment rate under 4 percent:** Cox, Jeff. "September Unemployment Rate Falls to 3.5%, a 50-Year Low, as Payrolls Rise by 136,000." *CNBC*. CNBC, October 4, 2019. https://www.cnbc.com/2019/10/04/jobs-report---september-2019.html.

203 **earned from *The New Yorker*:** Khatchadourian, Raffi. "The Doomsday Invention." *New Yorker*. New Yorker, November 23, 2015. https://www.newyorker.com/magazine/2015/11/23/doomsday-invention-artificial-intelligence-nick-bostrom.

203 **his 2014 bestselling book:** Bostrom, Nick. *Superintelligence: Paths, Dangers, Strategies*. Oxford, UK: Oxford University Press, 2014.

CHAPTER 7: THE LEADER OF THE FUTURE

209 **Theory X and Theory Y:** McGregor, Douglas. *The Human Side of Enterprise*. New York: McGraw-Hill, 1960.

209 **a Theory Z to explain Japan's economic success:** Ouchi, William G. *Theory Z: How American Business Can Meet the Japanese Challenge*. New York: Avon, 1993.

210 **founded in 1945:** "About ILR." ILR School, Cornell University. Accessed October 6, 2019. https://www.ilr.cornell.edu/about-ilr.

210 **long, cramped huts:** ILR, Cornell. "Cornell University's ILR School: The Early Years." YouTube, November 18, 2015. https://www.youtube.com/watch?v=ED1DZQj2dBQ.

212 **Henry said in 2018:** Ricau, Pierre-Yves. "A Silent Meeting Is Worth a Thousand Words." *Square Corner Blog*. Medium, September 4, 2018.

https://medium.com/square-corner-blog/a-silent-meeting-is-worth-a-thousand-words-2c7213b12fb6.

214 **a *New York Times* op-ed:** Grant, Adam. "What Straight-A Students Get Wrong." *New York Times.* New York Times, December 8, 2018. https://www.nytimes.com/2018/12/08/opinion/college-gpa-career-success.html?module=inline.

215 **Mark Zuckerberg donated $100 million:** Hensley-Clancy, Molly. "What Happened to the $100 Million Mark Zuckerberg Gave to Newark Schools?" *BuzzFeed News.* BuzzFeed News, October 8, 2015. https://www.buzzfeednews.com/article/mollyhensleyclancy/what-happened-to-zuckerbergs-100-million.

216 **a 2017 AP investigation found:** Flaccus, Gillian, and Geoff Mulvihill. "Amid Booming Economy, Homelessness Soars on US West Coast." *Associated Press.* AP News, November 9, 2017. https://apnews.com/d480434bbacd4b028ff13cd1e7cea155.

217 **Amazon has donated $130 million:** Feiner, Lauren. "Amazon Donates $8 Million to Fight Homelessness in HQ Cities Seattle and Arlington." *CNBC.* CNBC, June 11, 2019. https://www.cnbc.com/2019/06/11/amazon-donates-8-million-to-fight-homelessness-in-seattle-arlington.html.

217 **Anand Giridharadas, a leading voice in this movement:** Giridharadas, Anand. *Winners Take All.* New York: Random House, 2019.

217 **Amazon will still build an HQ2 in Virginia:** Feiner, Lauren. "Amazon Will Get Up to $2.2 Billion in Incentives for Bringing New Offices and Jobs to New York City, Northern Virginia and Nashville." *CNBC.* CNBC, November 13, 2018. https://www.cnbc.com/2018/11/13/amazon-tax-incentives-in-new-york-city-virginia-and-nashville.html.

218 **Amazon fought back:** Semuels, Alana. "How Amazon Helped Kill a Seattle Tax on Business." *Atlantic.* Atlantic Media Company, June 13, 2018. https://www.theatlantic.com/technology/archive/2018/06/how-amazon-helped-kill-a-seattle-tax-on-business/562736.

218 **tens of billions:** Honan, Mat, and Alex Kantrowitz. "Mark Zuckerberg Has Baby and Says He Will Give Away 99% of His Facebook Shares." *BuzzFeed News.* BuzzFeed News, December 1, 2015. https://www.buzzfeednews.com/article/mathonan/mark-zuckerberg-has-baby-and-says-he-will-give-away-99-of-hi.

219 **Amazon AI tool gone bad:** Dastin, Jeffrey. "Amazon Scraps Secret AI Recruiting Tool That Showed Bias Against Women." *Reuters*. Thomson Reuters, October 9, 2018. https://www.reuters.com/article/us-amazon-com-jobs-automation-insight/amazon-scraps-secret-ai-recruiting-tool-that-showed-bias-against-women-idUSKCN1MK08G.

224 **J. Robert Oppenheimer:** Ratcliffe, Susan. *Oxford Essential Quotations*. Oxford, UK: Oxford University Press, 2016.

226 **Twenty-five US federal agencies:** "NITAAC Solutions Showcase: Technatomy and UI Path." YouTube, March 29, 2019. https://youtu.be/IakpZK9q6ys.

INDEX

Note: Page numbers in *italics* refer to illustrations, charts, and graphs.

INDEX

INDEX

INDEX

INDEX